# Dino,
# Godzilla,
# and The Pigs

# Dino, Godzilla, and The Pigs

*My Life on
Our Missouri Hog Farm*

Mary Elizabeth Fricke

Copyright © 1993 by Mary Elizabeth Fricke
Photographs © 1993 by Mary Beth Meehan


All rights reserved.

Published by
Soho Press, Inc.
853 Broadway
New York NY 10003

Library of Congress Cataloging-in-Publication Data

Fricke, Mary Elizabeth, 1951–
Dino, Godzilla, and the pigs: my life on our Missouri hog farm/
Mary Elizabeth Fricke.
p.    cm.
ISBN 0-939149-96-6
1. Fricke, Mary Elizabeth, 1951–    .  2. Farm life—Missouri—
Hermann Region.  3. Family farms—Missouri—Hermann Region.
4. Swine—Missouri—Hermann Region.  5. Farmers—Missouri—
Hermann Region— Biography.  I. Title.
S417.F72A3   1993
636.4′0092—dc20

[B]                                                     93-17898
                                                          CIP

Manufactured in the United States
10  9  8  7  6  5  4  3  2  1
Book design and composition by
The Sarabande Press

*To the American farm woman
(the backbone of the family farm)*

My deepest and most sincere gratitude goes to my family, especially my husband, sons, and father-in-law; to "Paula," "Ted," "Pete," "JoAnne," and the memory of "Bonnie" and all who allowed themselves to become characters in this book.

A special thank-you is reserved for Susan Baugh, regional director of the International Women's Writing Guild, and to Patricia A. Bosh, former secretary of the Hermann Chapter of the Missouri Writers' Guild, for their endless encouragement and support . . . for being *my* backbone.

Last, but certainly not least, this book would not have been published without the frequent advice and understanding of an incredibly patient editor. Therefore, much gratitude also goes to Laura Hruska for allowing me the opportunity to prove myself as a writer.

# Dino,
# Godzilla,
# and The Pigs

# Chapter One

My eight-year-old son plays in the creek that divides our place from my husband's parents' property. Usually it is shallow and he can jump between the jutting rocks and catch crawdads, tadpoles, and frogs in the ankle-deep water. But several times after storms I have seen this docile creek become a roaring wave of life-threatening water capable of washing a pickup truck miles downstream.

The little creek that begins as a mere trickle on our land is nearly twenty feet deep and forty feet wide by the time it reaches the Missouri River at the lower end of the town of Hermann, some five miles from our home. We live in the state of Missouri, part of the region known as the heartland of America. It's a region that, I'm told, seems remote now, even foreign, to urban and suburban Americans who are the vast majority of the population. It's a region that has undergone its own changes, as progress has come to the farm in the last few decades. Yet there still are many medium-size and small farmers like us at work here (although the big agri-con-glomerates may outdo us in total production and profit). And though each of us, individually, contributes but a small amount to the life of this nation, taken together, we amount to a vital force in this country. Maybe not a flood but surely more than a trickle.

They predict that farms like ours and lives like mine will soon be mere memories. Maybe. If so, this book will show what it was like to live in this other United States in the last decade of the twentieth century. But it is my hope that the predictions will prove wrong, and that our farm and others like it will survive and endure for many years to come.

Missouri lies in the eastern half of the nation's farm belt. It is primarily rural, but heavily populated nonetheless. Hermann is located in the central part of the state about eighty miles west of St. Louis. Kansas City, one hundred and fifty miles to the west of us and St. Louis are our really big cities.

4

Many of Hermann's old buildings have been in continuous use since it was founded in the early 1800s, and some of the streets still do not have sidewalks. Its 2,754 inhabitants are mainly of German descent. It seems as if everyone is related to, or at least acquainted with, everyone else in Hermann.

Hermann's principal public buildings occupy strategic positions: The United Church of Christ and St. George's Catholic Church set on the highest hills to the northwest; the non-denominational house of worship is located further east, confronting those who enter Hermann by crossing the bridge over the Missouri River. This is also where the Gasconade County Court House and jail are located. The Baptist Church and the Methodist Church greet those who come to town on highways from the southwest. These public buildings convey a message: This is a law-abiding and God-fearing community.

In addition, Hermann has one small clothing store, two supermarkets, two catalog branch offices, a Ford dealer, five auto repair shops, a lumberyard, a Western Auto and a 7-Eleven, a landscape gardening center and an FTD florist, two veterinarians and several law offices, two photographers, some antique stores and hobby shops, several taverns and many more restaurants and motels and bed-and-breakfasts, two banks, a major office supplier, a library, a hospital staffed by four local physicians as well as a nursing-care center and a funeral home, a Catholic elementary school, public schools, the post office, a local newspaper and a police station. We also have four factories, and grain storage elevators belonging to several dealers. Hermann is home to the Veterans of Foreign

Wars, the Jaycees and the Knights of Columbus as well as the Eagles. The John Deere dealer is just down Route 100.

Hermann is surrounded for miles by one farm after another—crop and grain farms; hog, cattle, and even vegetable- and fruit-producing operations. The flat Missouri River bottom to the north and south and the Gasconade River to the southwest are responsible for the abundance of rich, arable land. This is beautiful country. The original German settlers chose it to remind them of their homeland. But agriculture is rivalled now by tourism. In recent years, the town, which has several flourishing wineries, has become a major tourist attraction for midwesterners, especially during the Maifest and the month-long Oktoberfest.

At my first sight of Hermann I, too, was impressed with the old-fashioned charm of the community. But having to make long drives whenever something has to be bought in a hurry makes charm pall. We live more than twenty miles from the nearest mall. Any time we run out of something, or when I need clothing or a household item, it's a half-day trip for me to drive back and forth. Recently there was a move to bring in a Wal-Mart and I was all in favor of this. But the tourism faction won out over the farmers and it was decided to keep Hermann as quaint as it was the day I drove here to help a friend and fell in love.

When I met Dennis Fricke I was twenty-nine years old, a recent widow with a six-year-old son. I had not lived on a farm since my early childhood, and although my stepfather, who made his living in the construction business, farmed on the

6

side, as a girl I had nothing whatever to do with farm work. My mother had been raised in a family of fifteen children of whom eight were boys, and she insisted that girls should not do "men's" work. So I was not prepared for life on a farm.

I had been married young the first time and had worked as a teacher-aide in our local parochial school for two years. We had a son, Jack, and then John, my husband, was disabled by diabetes. I cared for my invalid husband for six years while I babysat for nine children in our fourteen-by-fifty-six-foot mobile home. Then, in February of 1980, John died.

During the months of June and July of 1980 the entire Midwest underwent a record-breaking heat wave and dry spell, a precursor of the farm crisis that occurred later in the decade. It didn't take drought and excessive heat to make me feel crazy. Though I had loved him, when my husband died it was as if a burden was lifted from me. I felt almost light-headed. I stopped babysitting; I couldn't stay tied to that mobile home for another minute. I found a job as a sales clerk in a small department store fourteen miles away and investigated the possibility of attending college twenty miles still farther off to obtain a journalism degree.

My parents had always seemed to me a little too protective, especially where John and little Jack were concerned. Now they literally hovered. My stepdad could not understand why I didn't just close up the mobile home and move back in with him and Mom. But I had finally achieved some freedom and I was not going to give it up, no matter what.

Most of my friends were married, or in relationships; it

7

seemed as if we no longer had much in common. But at the same time that I was widowed a friend from my teens with whom I had worked in a shoe factory briefly was going through a sticky divorce. We began visiting; we had problems in common.

My friend had always had difficulties in picking men. It seemed to me that out of any group, she was certain to choose the sleaziest, most dishonest guy, the one most likely to cheat and abuse her. She was the daughter of an alcoholic father and a victim of child abuse, which seemed to me to be the cause of her lack of judgment. One very hot July night I drove eighteen miles from my home to hers after work to console her after her latest crisis. I found she had not eaten for twenty-four hours, but the only place she consented to go in order to get some food—in Hermann, a town filled with restaurants and drive-ins—was a tavern where she knew the proprietor.

It was after 10:00 P.M. and there we sat, at a table next to the bar, in a crowded tavern packed wall to wall with men. As I drank my Coke and urged her to eat her hamburger, she pointed out the guys standing at the bar she knew, offering to introduce me. I kept declining, becoming more and more uncomfortable with our situation. Each of her selections looked progressively less trustworthy. But then, as we sat there, a tall, lean, blond, clean-looking fellow wearing faded jeans and a white T-shirt walked in and made his way to the bar in slow easy strides. Several of the men there called to him; even at a distance his smile was warm and engaging. Between strains of song from the jukebox, his voice was audible, deep and rich.

I said to my friend, "If you want to introduce me to someone, why don't you introduce me to him?"

With disgust, she said, "He's just some dumb farmer."

That putdown was all I needed to hear since I come from a long line of "dumb" farmers. "I can't figure you at all," I told her. "You've been pointing out bums to me all evening. When someone finally comes in who's not only clean and attractive, but looks honest and damned proud of it, you act like he's dirt. You need to see a shrink."

After a half hour of coaxing, she finally agreed to introduce me to this guy. She waved to him. He came over to the table, leaned on it, and stared at her curiously.

She said, "Dennis, this is my friend, Mary Beth."

He spared me a glance, shrugged, and said, "That all you wanted?"

She nodded and he went back to the bar. Then he spun around and did a double take. This time when he looked at me he smiled and his smile reached all the way up into his blue eyes. I guess I smiled back. He says I did.

He came back to sit down beside me. And we talked and talked. Before I knew it, it was midnight and Dennis was saying, "You're not driving home alone over that twisting Highway 94 at this hour. It's too dangerous."

So he drove me home and returned the next day to take me back to Hermann to retrieve my car. And we went on from there.

In these months of 110- and 115-degree heat, Dennis and his father, who were then in business together, had watched several hundred acres of corn dry up and die before their eyes. They were facing the devastating prospect of having no corn to feed to their two hundred and fifty sows in farrow. I knew nothing of this and Dennis did not mention it. But one of the very first places he took me to see was the new thirty-two-sow farrowing house and attached nursery he was building. To me it was just a concrete-and-metal structure. I skipped up and down the recently poured concrete gutters like a child, with no idea what they were for.

I didn't understand anything about farm life then. I had fallen in love with a man, not his occupation. It made no difference to me whether he was a doctor, lawyer, ditch digger, or beggar. I was determined to spend the rest of my life with him. I didn't realize that when I married the man, I married the farm.

As a bride, my dowry was an eight-year-old dog, Mollie-Moo; a seven-year-old car; my six-year-old son, Jack; and a mobile home that was not yet fully paid up. We sold the mobile home and moved into the old house on Dennis's grandfather Hugo's homeplace.

For the first few years of my marriage, I was not actively involved with farm work. I worked part-time teaching at the Catholic elementary school in Hermann, and then after I gave birth to our son, Kevin, I stayed home to care for him. A year later, in 1983, we began to build our own home near the hill upon which Dennis's parents, Virgil and Margaret, live. Their house has been in the family since the early 1800s and it's a traditional farmhouse with two-foot-thick stone walls, two stories high, with a big farm kitchen where Margaret presides, and a comfortable front porch.

Our house, also sited near the top of a hill, faces southwest but it's a far cry from the traditional. It is termed an "earth-contact" house because it extends twenty-eight feet into the side of the hill. We considered going about twenty feet longer and building an even more protected "finished" earth-contact house but I objected. I didn't think I'd like to live underground so much. The house has a fifty-foot frontage, now covered with boards painted red, but someday it will be faced in brick. Our temporary roof is of corrugated white tin. Our house has just one story, as yet, but we hope to add on a second floor . . . someday.

We began construction by digging the hole and having a foundation and three concrete walls poured for us. We finished the front at ground level leaving two four-foot-square windows in the kitchen and master bedroom from which we have spectacular views over the valley. The man who poured the concrete for us advised against putting windows across the back of the house where the living room, bathroom, and boys' bedrooms are because they would need gutters poured around the outer frames and would therefore be costly and also would draw moisture and have a tendency to leak. I protested; I could not imagine living in rooms without windows—even windows with no view—but we took his advice on this. After insulating and putting up interior walls, we moved right in. The concrete floors were bare, the ceilings were open insulation, and we didn't even have rods to hang clothes on in the closets for another eight months, but we survived.

Now I wish we had taken the plunge and built a finished earth-contact house. I have gotten used to it and find that the house is both convenient and inexpensive to heat and cool. The boys are rarely in their bedrooms in the daytime, and we get good light in the living room through an arch that opens into the kitchen. We have no air-conditioning and we don't miss it: In summer we use a couple of fans that we run day and night to pull the cool air from the back of the house toward the front. In the daytime I draw our insulated drapes and shades and the indoor temperature stays between seventy and eighty no matter how hot it gets outside. Our only problem is that we have to restore the moisture that our wood-burning stove draws from the air in cold weather, and in summer I cannot use my clothes dryer unless it is a chilly or rainy day because it makes the house too hot and humid.

A few years after we moved into this house, Dennis and I bought some land of our own, one hundred and sixty acres, ninety of which are prime farmland. There was a house on the property as well as ten outbuildings: an old flatboard-sided barn on which the paint had faded into the grain of the wood; three granaries with raised wooden floors; two chicken houses (one for hens, one for baby chicks) complete with brooders and other chicken-raising paraphernalia; two garages, one double with dirt floors (I found two broken wooden wagon-wheel axles in there just after we purchased the place), the other so narrow nothing wider than a 1949 Plymouth could fit into it, with a selection of old license plates dating back to the 1950s tucked behind the bare studding; an outhouse whose roof is braced with even older license plates; and a washhouse in the corner of the yard with two rooms and a storm cellar beneath (which we fear to enter because the foundation is cracked and the walls may cave at any time). The house, two-story white frame, with six very large rooms, high ceilings, hardwood floors throughout, heavy carved woodwork, a cellar, and two porches, is located on a hill. From the front porch you can look out over the fertile fields and pastures and see all the lovely, gigantic old trees. It was built nearly a century ago and two generations lived in it until only two widows and a retarded son remained. The place was sold so the proceeds, put in trust, could provide for his final years.

We considered leaving our in-ground house and moving into this wonderful old place. But we would have had to insulate, put in central heat, remodel the kitchen and bathroom, and lower some of the ceilings for heating and cooling efficiency. So we eventually decided to leave the old house and immediate surroundings just as they were—private, secluded, and peaceful—and remain in our modern house.

We have since added another bedroom to our home, have put down brown indoor-outdoor carpet in the living room and white-and-gold linoleum in the kitchen and tiled all the ceilings. Our living space now is comparable to a double-width mobile home. We don't have enough storage room and there's just one bathroom; we don't have the charm of the older house, but economics dictated our decision. If we are to continue to farm, then this was the only choice we could make.

Our furnishings are eclectic: We have bought a few pieces as needed but most of our things have been handed down from various members of the family. I suppose this is where our farmhouse tradition is preserved, to the extent that it is. Our oak dining-room table and chairs and a huge china cabinet were inherited from Grandpa Hugo. We have a rocking chair that belonged to my stepfather's grandmother and dates back from before the Civil War. Long before I met Dennis I'd taken it apart and refinished it piece by piece and it's my pride and joy. My bedroom dresser is a huge cherry oak piece I remember seeing in my grandmother's bedroom. We have an antique jade-based lamp and an incomplete spinning wheel that I confiscated from Jack's grandmother's attic. But I guess our most prized possession is the hard-back chair that was Dennis's grandfather's that Dennis sits on when he works at his desk. While Grandpa Hugo was alive no one else would have dared to set in it. The rest of our home furnishings are make-do.

What I do have is a glorious view over the valley from my kitchen window. In spring I see flowering buds of sumac, dogwood, oak, and maple against the ever-present deep green of the cedars. In summer I see the dense shadows of heavily laden trees in whose shade the wildlife takes refuge from the stifling heat. In fall, bright orange, red, and brown leaves blow through the sky and flutter onto the ground. In winter, the prevailing color is brown, except when it snows.

A ten-acre field at the base of the valley separates us from Dennis's parents' home. Other houses, far and near, dot the hillsides. There is a packing house about a mile down the road that draws a good deal of traffic during the day. But for the most part it is quiet here. The only sounds to be heard most of the time are the humming of farm machinery in the distance, the lowing of cattle, and the protesting squeal of a hog as our fifteen hundred–head herd shifts itself into a new alignment.

I can tell the sound of my husband's pickup truck or my father-in-law's or of either of the farm trucks when they start up. I know without looking which tractor is being used, and when my husband is grinding feed. I can hear the school bus a mile away and I know the sound of the feed truck when it comes to make a delivery. On summer evenings, the whippoorwills and the katydids among the trees serenade us. Toward daybreak I am often awakened by the yip of coyotes and the neighborhood dogs' answering howl. We sometimes hear a screech owl nearby, though we have not heard one recently. After dark we often sit in the yard barbecuing a late supper, and as we watch the fireflies under the stars, not a car passes.

There is scenic beauty on our farm and tranquility too, but there is one sense that farm life does not always thrill. While I doubt anyone will dispute the fact that there is no sweeter smell than that of fresh-cut flowers or newly mown hay, or even the fresh clean scent of laundry left upon the line to dry in the breeze, there are many other kinds of smells in the country. On hot summer days when the humidity is very high, the gases in the lagoon into which the wastes from the hog buildings drain form a cloud above the water and emit an odor that can be smelled for miles around. And I find no smell more offensive than that of rotting feed or corn, an odor caused by mold. Indeed, this mold expels a gas that can kill upon prolonged exposure. And this is by no means the only hazard of farm work.

Farming is a very dangerous occupation. In the fields we must not only be careful of the powerful machinery we use but also of the chemicals needed to produce large crops, some of which are saturated into the seed by the manufacturer and some of which we spray to kill weeds. Then there is our exposure not only to the sun but to dirt, grit, and oil: It is not unusual for one of us to take several baths in an attempt to rid ourselves of the stench that clings to our hair, our skin, and in our nostrils after a day spent working either with the hogs or in the fields. But sometimes I think that stress resulting from our inability to control so much that our lives, happiness, and prosperity depend upon is our worst health hazard. I think this is the prime factor that has driven so many families out of farming.

I have watched with dismay the end of many small family farms and the departure of the farm families who used to live and work there. Today, we are the caretakers for several who have left farming as a livelihood and a way of life. In addition to our own acreage we cultivate twenty separate properties owned by different individuals. The properties nearest our home and the ones that border the Gasconade River are all located on moderately traveled county roads. On those roads are attractive homes that are still lived in all year-round. But the houses on some of the remote hill properties are now used only on weekends, usually during deer- and turkey-hunting seasons. And some of the tracts of land we farm have houses on them that have not been lived in for years. You see the tracks of rabbit, squirrel, and deer among the waist-high weeds going right up to the steps of the houses, sometimes even inside on the dirt floors of barns and other outbuildings. Most of these properties belong to elderly people who have moved into town or to estates owned by relatives of a deceased farmer. We know most of the owners and try to care for their land as they would had they still been up to it.

Sometimes while I'm waiting for my husband in our travels from farm to farm I can hear over the steady whir of the tractor or combine motor the sound of birds and the step of an animal watching, hidden among the wood or brush that has grown up since the family moved away, or sometimes just the rush of a gentle wind through the trees and bushes. The smells are of fresh earth, wild grasses, and flowers. I can see, past the weeds and other signs of decay, a barn with its faded flatboard siding (once painted red, now grayish brown), its doors thrown wide. A forgotten plowshare sets just inside the door. An old rotting wagon on three wheels stands back in one corner. We piled extra seed sacks on that wagon one year to get them out of the rain and that night the coons tore them to smithereens. Remnants of the sacks are still there, as well as a three-quarter-inch-diameter horsehair rope once used to open the window above the loft through which the hay bales were lifted. Among the trees just off the drive stands what is left of a split-rail fence surrounding the rusting tin-roofed huts where the hogs used to wallow in mud and shaded comfort. Just over the rise, out of sight beyond the pasture, lies the machinery graveyard: an old pull-type combine; the frame of an iron-wheeled tractor; a rusted, flat-tired model-A truck with no windows and no seat either because mice and other rodents have stolen its stuffing for their own beds.

No matter which property we travel to, a small creek—now more like a ditch—separates the house from the farm sheds. In spring these creeks are usually filled with water. In the fall they dry out and sandy bottoms show. But a couple of places have real creeks with rock floors and springs, great places for kids to swim and play in.

Next to white frame or stone houses lie the remains of gardens. Some still have asparagus growing each spring. One has a grape arbor, with grapes hanging in sparse bunches from its branches each August. Weeds have taken over; fences are falling down. Still, I can see the rows and hills where pumpkins, potatoes, tomatoes, and green beans grew.

In the yards are flower beds and climbing roses entwined and left to latch onto trees, clothes line poles—whatever they can find to grip. Daffodils rise in the driveways come spring and irises have taken over the drainage ditches beyond the kitchens. Wild violets blanket the shady spots in the yards.

The once white houses are now dull gray, their roofs covered with rust instead of shining silvery tin. Bricks have fallen from chimneys, and even though the cisterns are concrete they look unsafe. We rarely venture inside these houses. Most are locked and boarded shut, windows blank, shades drawn.

The farm where Dennis grew up until he was sixteen is one of the properties we cultivate now. Each time we return to the place of his childhood, Dennis begins to reminisce. He remembers catching crawdads in the creek with his sisters, the tire swings that hung in the trees, the firewood that he carried from the shed to the woodbox that still stands on the front porch, putting hay bales up into the barn loft in the summer. It is not hard to imagine children doing these things on any of the properties that we farm, to hear their laughter as they splash in the creeks and their squeals of delight as they swing on the ropes hanging from barn lofts or in the tires hanging from the trees. But there are no other families here now—only ours.

# Chapter Two

In addition to our house there are several other buildings on our place. An addition to the original barn, built on log stilts upon a log base, still stands at the foot of the hill; we use it as a storage shed. We also have a newer shed with an open side and a tin roof where we keep vehicles and farm machinery: a car, a pickup, a two-ton truck with a hoist bed and racks, four tractors, two discs, a drill, a combine, a planter and auxiliary machinery such as the feed grinder, auger wagon, bush hog (weed mower), mallboard plows, Kuker sprayer, and of course, the "honey" wagon. Even a small operation like ours needs a lot of capital equipment.

Then there are the two large buildings in which we raise hogs: the farrow building, a structure one hundred and eighty feet long by twenty-four feet wide that is divided into three rooms, and the finishing building, which is two hundred feet in length. In addition, there are outdoor sow pens. The hog-raising operation is located downhill from our house, with a lagoon (and an overflow lagoon) into which the wastes from the farrowing and fattening buildings drain located lower still. And then we have several large Baughman bins in each of which up to five thousand bushels of grain can be stored as well as several feed bins that hold between six and twelve tons each.

We run a mixed farming business: In addition to our hogs we raise field crops, wheat, corn, and soybeans. I don't think we could manage financially if we had to buy grain to feed our herd of hogs. A good fifty percent of our income comes from the grain crops that we grow on our own and on rented land. Hog-raising operations the size of ours are standard for the *small* farmer today. The days of a farmer keeping two or three sows and a boar in the pen behind the barn are gone.

Many hog farmers operate on a much larger scale. They are usually "consignment" farmers and employ a large work force to care for sows or "feeder pigs" (piglets born elsewhere) under a contract pursuant to which feed and other necessities are supplied by the piglets' owner. The consignment farmer is often paid a flat fee per hog, although the terms of each contract vary and sometimes he may share in the profits, if any. Other large-scale hog farmers own their own hogs but finance their operations by contracting to deliver them at a future date for a set price. If the hogs don't sell on that date for that price, the farmer has to make up the difference from his own pocket.

We are currently operating at a one-hundred-to-one-hundred-and-twenty-sow capacity but we have had as many as two hundred and fifty sows at our operating peak. Since a herd is usually measured by the number of breeding sows, this doesn't account for boars, gilts, or piglets. Small as we are, we have a lot of pigs on our farm. But in 1986 our herd's size dropped radically to less than a hundred sows due to one of the natural disasters that farmers are prey to. It was in that year and for the same reason—necessity born of that disaster—that I began the transition from farmer's wife to farmer.

Dennis has always been ambitious, and in the spring of 1986 he rented several hundred acres along the banks of the Gasconade River. It had seemed to be a good move; hog prices were at a record high, and crops appeared to be the best ever. And then, in September, it began to rain.

It rained almost continually for three weeks; we had nine inches in a single week of showers. In mid-October both the Gasconade and the Missouri rivers were in flood. We lost more than four hundred acres of crops, watching them drown beneath the relentless rains and twelve-foot crests of the rushing rivers.

All of central Missouri was designated a disaster area. The town of Hermann was cut off from civilization for several days;

food and emergency supplies were flown in by helicopter. Weeks passed before some businesses opened their doors again. In little Rhineland, on the other side of the river and more than a mile inland, floodwater reached the second stories of buildings. In Treloar, eighteen miles away, the levee broke on the edge of the farm where my stepfather and his father had been born and raised. When the water receded, his wheat field was buried beneath twenty feet of pure white sand.

The corn and soybeans we combined from the fields in the spring of '87 were unsaleable. What the river had not washed away it left littered with dirt and debris. We used some of the crop for feed, but the animals were reluctant to eat it and did not fatten well. Whenever Dennis ground corn into feed, a black cloud of dirt rose high above the feed bin and could be seen for miles. My beloved dog, Mollie-Moo, died. (We replaced her with an English shepherd.) And then my son Jack, who was now twelve, had to be hospitalized and the diagnosis was serious. He had Type I diabetes, the disease that had killed his father before he was thirty-two. I guess it was understandable that Dennis's blood pressure soared and he had to start taking medication.

When our hired hand sought a better-paying job elsewhere, Dennis proposed that, instead of my seeking an outside job for which I would earn a salary, I begin to work on the farm with him.

When I write of this idea now, it doesn't seem so unusual. But it seemed so to me at the time. Margaret, Dennis's mother, had been married to a farmer and raised three children, but the household and garden were strictly her domain. The rest of the place was her husband Virgil's responsibility. And for my own mother, work was still divided by gender. We knew few couples who were running or attempting to run a farming operation in tandem. Most who needed the money followed the traditional path: The wife would take a job—part- or full-time—in town.

But to us it seemed sensible to keep the farm work in our family. Jack took on the daily job of feeding and cleaning the animals in the farrowing, nursery, and finishing buildings, and I became the extra hand, involved in every other aspect of farm work.

The first job I undertook all by myself was in the farrowing room. I was to cut the teeth and tails and give shots to about one hundred three-day-old piglets. Pigs raised on wire or slat flooring cannot get the iron they would otherwise root out of the earth so they must be given iron shots by the time they are three days old to prevent anemia. And their teeth and tails must be trimmed because otherwise they will bite off each other's tails in play or do serious damage to one another with their tusks. They weight only three to five pounds, but they wriggle and squirm and they relieve themselves out of fear, and they run, run, run. And while all of this is going on, their six-hundred-pound mothers are right there, looming. Today I can laugh about it, but then I was nearly overcome with fright.

I had grown up near hogs but my stepfather's animals, sows and boars alike, rarely weighed more than three hundred pounds. Some of our sows are double that size. Besides, my stepfather had never allowed us to go near the hog pens. And many times we had heard of him being chased from a pen by a sow who had just given birth.

It is true that over the previous five years I had occasionally fed the sows and helped to move them to and from buildings, but I had remained outside while the men went in and penned them up. When I'd fed them, I had been separated from their huge round heads and nuzzling snouts by a thick three-foot-high steel gate. And when you move sows, the rule is to remain in back of them, otherwise they will turn or back up, but will not walk straight ahead. No matter how large they are, somehow they look far more ridiculous than frightening from the rear.

I had also helped to move piglets from the farrow room to the nursery. But the sows had been taken out of the pens and housed elsewhere beforehand and the little ones rarely weighed more than twenty pounds. It was fun for little Kevin and me to try to catch them. But now I was supposed to reach inside the pen of a new mother and to handle her piglets— naturally, I was terrified.

In the farrowing building a sow is housed in a crate (approximately eight feet long and three feet wide) made of steel bars. She can stand, sit, or lie but is not supposed to be able to turn around. Still, I have seen more than one manage this feat with little difficulty. A sow cannot jump out of a crate; the bars on top prevent this. The crate, in turn, is surrounded by a walled pen six to eight feet square and two feet high. The piglets are free to roam within the pen until they are weaned, then they are moved. We used to shift them into the nursery for three to four weeks until they weighed about one hundred pounds, at which point they would be moved into large pens (approximately fifteen by thirty feet) in the finishing building. Today, we leave them in the farrowing room a little longer, then move them directly to the finishing building.

Feeders and waterers are attached to the walls of all pens so the pigs may feed and drink at will. The pens stand on steel-wire or wooden-slat floors so that wastes drop into gutters (or pits) beneath the buildings that then empty into a nearby lagoon. Cleanup is a minimal job as the pens need only to be hosed down regularly to keep them sanitary.

That first day, I patiently waited behind each sow for her piglets to come within my reach. There is nothing wrong with this method, but it is time-consuming.

It didn't take me very long to discover there is a technique to be mastered. Catch the piglets when they are nursing or sleeping: If they are asleep they don't squeal, and if they are nursing, the sow is lying down. Otherwise, the squeals of the first piglet will awaken every animal for six pens down and sows are extremely quick when agitated. They are not evil or vengeful but a sow will attempt to protect her offspring just as the boars try to protect their herd. (Dennis laughs at the idea that a boar will protect anything. But I ask you, what are they doing when they fight other boars if not protecting the sows they consider theirs?)

I emerged from the building smeared with iron injectant, blood, and pig excretions along with plain sweat. My hair was full of cobwebs, my eyes were dazzled by the sun. A job that should have taken about an hour had taken me four.

Dennis and Virgil were amused, though they didn't say anything. But that didn't matter. I was so proud of myself. I had conquered my fear of sows: I had proven that I could get the job done.

That first spring when I began to do real farm work I had to learn to drive the tractor. Dennis had a hard time convincing me that I could not work on the farm without mastering this skill but he finally persuaded me. He started me out on one of our largest tractors, our John Deere 4040, for safety's sake, since most accidents occur in smaller tractors that do not have cabs, dual wheels, or split-second controls. But I have to admit that the size of the piece of machinery that he proposed entrusting to me was daunting.

Thinking back now, I am surprised that I learned as easily as I did. I don't really like to drive. I have never learned how to operate a standard shift. I am one of those people who prefers to use the natural transportation God gave me . . . my feet. And tractors have really tricky shifts: There are sixteen on the one I learned on.

"You don't have to know every single shift on the damned tractor," Dennis would growl. "All you need to remember is the one I tell you to use."

And so I found myself reluctantly seated behind the steering wheel, where I soon discovered for myself that, large as these pieces of machinery are, they are relatively simple to operate.

For safety reasons the steering and brakes are made to operate with very slight pressure. I found there is nothing like the thrill of sitting high on the seat of a tractor while traveling over hills and meadows, and there is enormous satisfaction in watching the earth fold over the disc blades behind the tractor as we prepare the ground to grow acres of wheat, corn, or soybeans, doing our part to grow the food that helps to feed the world.

When I first began to use it, our old disc seemed like such a monster, I started calling it Godzilla. Maybe giving it a nickname helped to tame it for me. It had a fourteen foot span and was about twelve feet long, with four rows of blades. (Our new disc, which I named King Kong, has a span of twenty-five feet, twice as wide as the tractor that pulls it, and weighs about 10,000 pounds. It has its own hydraulic unit on a big green metal frame from which extend two arms. Each arm weighs nearly five hundred pounds.)

Later my first week discing with Godzilla I was working on property that borders the Gasconade River. The field was surrounded by a fence enclosing some thirty head of a neighbor's cattle. It was nearly 10:00 P.M., time to quit for the day, but I had to refuel the tractor first. Virgil's pickup carrying the portable fuel tank stood on the opposite side of the fence. Refueling was simple but as I began to drive off I forgot that I had Godzilla—three feet wider than the tractor on each side—hitched behind the tractor. It was the expression of horror on my father-in-law's face that stopped me; I thought I had killed someone. Ten feet of fence were torn up before I realized what had happened.

When I saw what I'd done, I sat down on the disc wheel and cried. Virgil patted my shoulder consolingly and told me that he had torn out a few fences in his day too. He repaired as much of the damage as he could, and when the fence seemed safe enough for the night he drove off.

Dennis had been planting in another field but halfway home I had to confess. He was furious at me, and his father came in for a few choice words too.

The next day, as he repaired the fence properly, while I made ineffectual efforts to be of help, Dennis asked me in the most irritating way, "Have you at least learned your lesson?"

But about four hours later I managed to hook our neighbor's gate with the field cultivator and to drag it a good six feet before I noticed a thing.

I've learned a bit since then, but I still make mistakes and Dennis still points them out to me. Sometimes it's hard to be married to your boss. There's no one you can complain about him to, after work.

# Chapter Three

In spring after the earth has been prepared with the disc we plant the seeds, then carefully watch them grow. Within a week after each field has been planted, Dennis begins to check the rows for sprouts. When the plants are three inches high he sprays the ground between the rows for weeds. Within a few days we walk the rows of every sprayed field to make certain the spray took hold and that it was not so strong as to burn the tender leaves of the young corn or soybean plants. Then we set back and watch our crops grow, checking the fields often, praying for the right amount of rain at the crucial time: when the corn tassels, when the soybeans bloom.

Healthy corn stalks are a rich dark green. But if there is too much moisture in the soil the centers of the stalk and of the leaves are a lighter color, verging on dull yellow if the ground is really wet. If the crops have received too much insecticide or herbicide, they turn brown and the leaves droop and curl under. When the leaves curl up and inward and the cornstalk drains of color slowly, turning almost white, the plant needs water. But Dennis has difficulty discerning these telltale color clues. He is color-blind.

I first became aware of Dennis's problem when we were on our honeymoon and he asked me to watch the traffic lights for him. Until then I had never noticed the vast difference in the positions of the red and green portions of traffic lights in various towns and cities. Unfortunately, his handicap is just as serious in farm work. If he is checking the crops alone he must walk among the plants and feel the texture of the leaves and examine the ground for signs of moisture. This is not only time consuming. If the change in the texture of the leaves is discernible manually it may be too late to salvage the plant. Now I often ride with Dennis to tell him exactly what the color of the plants is and how many of them are affected.

By midsummer, ears of corn or pods of beans should begin to form. The size and circumference of the ears, and how well the kernels grow out to the tip of the ear, will determine whether our corn crop will be successful when we harvest. The number of pods per plant and the fullness and firmness of the beans determine the success of the soybean harvest. These are the factors our income depends, the determinants of profit or bankruptcy. Ironically, in "good years" when all the farmers have bumper crops, grain prices drop so low that sometimes the crop is not worth selling. We may end up storing it in our own bins or in commercial bins rented from a nearby feed-and-grain service. In "bad years" when due to inclement weather there are no crops, no fruit of the harvest, no income, the foresighted farmer has his grain stored from previous years to fall back on. And then the price per bushel will be very profitable.

Our wheat crop is much the same except that we plant winter wheat, sowing in the fall and harvesting in June, to bring in some income in early summer and even out our cash flow.

In the stifling heat of August, healthy corn abruptly turns from rich, dark green to golden brown. When the ears are full they point outward like the ears on a donkey. By the first frost the soybean plants have dried and ripened and lost all their leaves. Looking out over a field of ripe soybeans you see stalk after brown stalk of fuzzy little golden brown pods the length of each stem.

I like the brisk winds of fall. Sometimes they blow in storms but most of the time they offer a continual refreshment to the senses. For as each crop has its distinctive appearance when ripe, each also has its own fragrance. Corn and wheat are both sweet and "sundried," while soybeans emit a pungent, musky odor that makes you gasp. Sometimes in early fall there are hay fields still to be cut and baled. The smell of new-mown hay is always a joy. And when the crops are finished and the land turned over to expose the roots of the dead stalks, there is once again the aroma of fresh, clean earth.

Some people await the turning color of the leaves in fall, but on the farm we wait eagerly for our crops to ripen. When the combine is driven into the corn field for the first time we watch closely, holding our breath in anticipation as the bin fills with that first load and the huge combine pulls alongside the truck to dump the brilliant yellow kernels from the auger onto the waiting truck bed. My sons, and nieces and nephews too, are always eager to help unload the corn from the trucks into the storage bins. The moment they hear the truck pulling into the driveway, they run to the bin. There they wait until the rear of the truck bed stops just over the hopper of the sixty-foot-long auger, which empties through a hole in the roof of the bin. Then up they climb, over the sides of the wooden truck bed, swift as squirrels and as noisy. With whoops and hollers they drive headfirst into the three hundred–odd bushels of corn that the truck holds. Diving, jumping, tumbling, again and again. When their shoes and socks are filled with corn kernels, off they come, to be tossed over the side of the truck. Kernels caught in pockets and creases turn up later on the bathroom floor, the living-room carpet—anywhere and everywhere. The kids are not allowed to remain in the truck bed as it unloads for fear of accidents. A scuffle could cause a fall; suction could draw the corn and a child down into the auger; there is a deadly possibility of suffocation. But they are allowed to play before unloading begins and they look forward to this rite.

Wheat chaff is sticky and soybean pods carry a kind of fuzz that is not easily separated in the thresher. So trucks loaded with soybeans or wheat are not only dusty and dirty, but they cause tremendous itchiness—no one wants to play in them. We do not unload either crop into our own storage bins. We sell them straight off the truck or store them at a local elevator.

While corn is usually the first crop to ripen, it will also stand well into the winter months even in the heaviest weather, so it may not be the first crop to be harvested. Soybeans tend to fall to the ground. If the weather is too dry the soybean pods will burst, casting the beans onto the ground where the sickle on the front of the combine cannot pick them up, and if the weather is wet and rainy the moisture will add weight to the plant so it cannot stand. Bean fields hold more moisture than corn fields. A one-inch accumulation of rainfall upon a field of corn will soak into leaves and stalks more than into the ground. But rain on a field of ripe soybeans will not only weigh the plants down, it will soak directly into the ground, rendering it impassable to heavy equipment for days. Wind dries the land, but also tears down the sensitive soybean plants. Since soybeans are a cash crop, the better the condition when sold, the larger our profit. Since much of our corn is stored and ground for feed, perfection is not crucial. For these reasons, we often harvest our soybeans first.

But corn that has retained too much moisture in stalk or ear must be dried; it should not be sold with a moisture factor of higher than fourteen percent. Otherwise the dealer will dock money from the market price, which can lead to financial disaster, especially if that price is low to start with. So one of the familiar sounds of autumn is that of the grain dryer. Large electric fans behind the bins blow air up into them through air holes in the bin floors. The motors are very large and so noisy that sometimes they can be heard for miles. When ours are first turned on it takes days for me to get used to the continual roar. They run twenty-four hours a day; we hear them in our dreams. But they act as a form of communication. When neighbors a mile or more away begin to harvest, we know it because of the noise of the fans beneath their bins. And if there is an electric power outage, the sudden silence is deafening.

Before he started school, Kevin usually rode in the cab of the 4040 John Deere with me. When, finally, I mastered this tractor I became so fond of it I called it Gentle Ben. It has a wide seat and an arm on the left side that can be lowered. When we were on flat terrain Kevin sometimes sat there, on the arm. Most of the time he would ride behind me. I am so short that I have to move the seat as far forward as it will go so I can reach the brake and clutch pedals. This leaves a snug little space between the seat and the back of the cab. I would latch the rear window tightly and pile in pillows and padding so Kevin could sit back there for hours. He would play sing along with the radio or watch the disc for me, or most often, nap. This is not a practice recommended by the manufacturer or the dealer. Tractors are intended to be one-passenger vehicles. I would never put a child on the fender of a tractor; I don't like to sit there myself when the tractor is moving. But the spot behind the seat proved safe enough and when there was no one to babysit Kevin, I didn't have much choice.

After Kevin began kindergarten, when I was in the fields in the morning I'd have to rush to meet the school bus that dropped him off at noon. Too many times I was a few minutes late. So I began to try to stay home mornings and then we'd go out to the fields together.

Kevin was with me when we began discing the cockleburs out of a field as Dennis combined the soybeans from it. The cockleburs stood higher than the cab of the tractor and were so thick in some areas that they could not even be combined. If we had been able to sell that troublesome weed, we could have made a fortune. We attacked the cockleburs, first from one direction, then another. Some areas had to be disced three and four times to cut the stems of the stubborn weed. As the hood of the tractor struck the plants, the sticky, spiny seeds flew in all directions like little black hornets. They hit the cab with a popping sound, like hail. They hung from the windshield wipers. Kevin and I were giggling at the mess.

Then we saw the rabbits.

The poor little creatures—there seemed to be hundreds— had made their homes in burrows beneath the thick cocklebur underbrush and to our horror some were not able to escape the sharp steel blades of the disc.

Farmers learn from childhood on to respect, protect, and preserve every aspect of nature—except, of course, for cockleburs. But even our improved technology seems unable to solve this conflict between nature and machinery. We had heard horror stories of little fawns accidentally caught in the grain reel of a combine. Mice, moles, rabbits, turtles, snakes, groundhogs—all are prey to farm field equipment. These animals hide in hay, wheat, or weeds or make their homes beneath the soil where they are invisible. On a return trip I would find them lying in a previously worked disc row. For the sake of human safety, combines, tractors, discs, and hay bines are manufactured with safety shields, warning signals, and controls that respond to the touch in a split second; I wish it were possible to give the same attention to animal safety.

This may seem hard to equate with the fact that we are engaged in raising animals to be butchered. It's not really. No farmer can abide waste. The butcher hogs we raise go to feed people. But wanton slaughter is an offense against man and God.


It took me three years of doing fieldwork before I was willing to try to drive the big 4320 John Deere tractor. (Because of its awesome power, Jack named it Goliath.) I had to learn how to push the seat forward so that I would be able to reach the gears. The catch is just beneath the front of the driver's seat. With his long arms, Dennis can just reach down to adjust the seat to go all the way back. I have to stand on the ground on tiptoe, stretch to reach over around the clutch and over the standing platform, and push the catch in to adjust the seat to as close as it will go. Before I figured this out, if Dennis wasn't there to fix it for me, I ended up almost falling off the seat when I changed gears or else I was almost throttled by the steering wheel. Once I consented to attempt to use it, Dennis insisted that I drive in fifth gear in order to break up the soil with more speed and force. I didn't feel safe above third gear, which is much slower, so we argued a lot about that. It's still hard for me to get on and off this big tractor because I am short and it is built up so high that to reach the first step I have to lift my left foot level with my breasts, grasp the handle on the hood with my left hand and the one on the fender with my right, and heave myself up. One of these days, I've warned Dennis, I will injure myself in some vital part of my anatomy performing this routine, and then he's going to be sorry!

# Chapter Four

We acquired a new member of the family or, more accurately, after we made several attempts to find him another home, we found out that he had adopted us. He is a classic mutt, weighing about twenty pounds, with the short hair, coloration, size, and markings of a beagle; the long body and short legs of a dachshund; terrier ears; huge soulful brown eyes; a short bewhiskered black snout; and the bark of a St. Bernard varied by the wail of a coonhound. His disposition is that of a playful, pleasant Collie-shepherd and he is as stubborn as an old Dutchman.

The boys named him Wiener Schnitzel but the Schnitzel was soon dropped and we call him Wiener. To our surprise, Sally, our black-and-white, amber-eyed English shepherd, made little protest. No other dog is allowed two feet off the county road onto our place, but she allowed him to stay. Wiener loves to run. At first, he would follow the car and the pickups everywhere, running alongside at ten or fifteen miles an hour. This daily exercise has made him solid muscle from nose tip to tail, a wonder to all who see his paces.

The first time he heard gunfire—Dennis was doing a little target practice—Weiner came running, showing no fear. I wanted to see if perhaps he had been trained to track squirrel and rabbit so when he followed me into the woods I didn't protest (although I doubt if he would have responded had I ordered him to go home). But the only squirrels he yipped at were already so high in the trees you could barely see them. He did chase rabbits, but it looked as if he was trying to tree them too. We decided that Wiener was best at hunting mice. But he has turned out to be good company for me whenever I have to wait for Dennis out in the woods, which frequently occurs in hunting season.

I hadn't been the new hired hand for long before Dennis started complaining: I wasn't giving him enough hours. It seemed to him that I was spending entirely too much of my time working in the house or on other chores and too little helping him in the fields. It didn't matter that I tended a fairly large garden from which I canned the produce—corn, toma-toes, pickles, and potatoes. Or that I painted the front and trim of the house as well as Kevin's swing set. Or that I was trying to finish the manuscript of a romance novel and sell a historical novel that I'd written. (This has long been my ambition—to be a published novelist.) Or that, in between everything else, I had to do the farm bills and clean the house. And of course, there were the Confraternity of Christian Doctrine (CCD) classes that I had begun to teach when Jack was a little boy and that I continued to teach, becoming coordinator of the pre-school and elementary programs. None of this counted, as far as Dennis was concerned. He dared me to keep track of the hours I devoted to farm work, sure it would prove that he was being short-shifted. So for the month of October 1988 I wrote my schedule down: It proved I was doing about eight hours a day of farm work and everything else besides, which quieted him down for a while. However, I can understand why he chose that time to complain.

October is one of our busier months of the year on the farm and Dennis needed all the help he could get and then some. Farmers put in a long working day out of doors, and then office work consumes almost as much again of our time. Most mornings Dennis rises by five o'clock and spends an hour or so studying the incoming reports on the DTN screen and on television, as well as other reports or data that has accumulated. He must spend hours studying livestock and grain market reports, figuring sales volume and grosses and nets in his head. He talks over the telephone to feed, seed, and other marketing people. He plans our planting and breeding strategy months ahead. We deal with most of the book work together. Dennis handles all the ordering of feed, seed, and equipment and sees to equipment repairs when problems arise. I write out letters and reports and pay the bills for house and farm, which means I also juggle to keep the checkbook in balance. And, in addition to this constant indoor work, at planting and harvest times Dennis spends long hours out in the fields as well, and in our hog operations.

But I can only stretch myself so far—especially if I am being asked to make an extra effort to free Dennis from farm work in time for hunting. Or what's worse, if I'm asked to leave my other tasks to help with the hunt.

Most years Dennis has been lucky enough to finish the fieldwork—crops combined out, the majority of the land plowed or disced at least once in preparation for spring—before the week of November known as deer season. At one time Dennis was a dedicated hunter and trapper all year-round, not so much for the money the furs brought as for the lean, wild meat that is a staple of our diet. Hunting also helps us control a number of animals that are potentially dangerous to our crops. But over the years, as his work load has increased, he's had less time for hunting. Except for deer season.

Each year I see his eagerness increase as deer season nears. His shoulders lift and the spring returns to his step. It is not just the prospect of the hunt; it is a time when he will be alone in the woods, part of nature, that he looks forward to.

Many times when I have accompanied him I have been

amazed at his ability to sit still for hours, not moving a muscle, scarcely breathing. But he is alert and watchful; he sees every leaf sway, every twig move, every bird, animal, and plant. No sound escapes him. No kidding! And this is a man who claims he hasn't heard a word I've said half the time. But in the woods he hears everything. For one week out of the year, Dennis is in his element and I would not deprive him of this time for anything on earth.

On the other hand, although Dennis has tried to make a hunter out of me, he has failed. I accompany him now and then during deer season, if I can. I have carried a gun into the woods and participated in drives to discover where the deer are hiding. But the truth is, I don't insist Dennis learn to sew or crochet, or even read the things I write, and I don't feel that he has the right to drag me into the woods with a gun in my hands and insist that I hunt. Unfortunately, I haven't been able to persuade Dennis to see it that way, perhaps because the first time I went out with a gun in order to prove that I couldn't hit the broad side of a barn, I brought down an eight-point buck with one shot.

But I am impressed into service on deer drives. Deer hide in the woods and have to be driven out; this requires six to eight hunters who know the terrain and can move quickly. The hunters wait at strategic points to shoot the deer as they emerge from the woods. I know animal rights advocates will say this is unfair but I can attest that when the deer come running out at full speed only the very best hunter will make a kill, and those caught in the path of the deer are in danger.

Besides, we don't hunt for the thrill of it: Venison is a staple of our diet.

Our local conservation agency reports the number of deer killed in our county. Each year, the number is lower in our area, though it remains high to the south of us. We are beginning to wonder where the deer have gone and to worry. Dennis, his family, and friends will not kill young, undersized animals, or gravid does. But others are not so careful, and the herds are thinning out.

Last fall, three or four days into the hunting season, no one in the valley had even seen a deer. So Dennis decided to drive the woods, not to kill but just to see if there were any deer in hiding. We had finished the field and crop work; the animals had been tended first thing in the morning and would not need care again till late in the evening. Kevin and Jack were in school. So Dennis figured that my time was free and at his disposal.

Each morning Dennis and his father drove a pickup truck to the opposite side of the ridge behind my in-laws' house while I drove down to the farthest hog pen on the near side of our ridge and sat down to wait for ten or fifteen minutes. I'd have Wiener for company as I watched the second hand on Dennis's wristwatch that I had borrowed tick off the time with agonizing slowness. Then, at the appointed minute, I would begin to walk back across the creek, through the hay field, into the woods, to the far end of the valley. And then I would climb the ridge, nearly two hundred feet straight up. I wore gloves but they were little protection against the sharp rocks that scraped my fingers. I carried a walking stick but still stumbled into holes and deep crevices deceptively filled to ground level with crumpled wet leaves. Each day I set off climbing a different path so there was no marking my way. I would tell Wiener to go home, but he never obeyed. Instead, he would disappear for fifteen minutes at a time and then reappear at the top of the ridge to yap at me as I struggled below.

When I managed to reach the ridge top I had to crawl through the thickest cedar stands and brush patches I'd ever seen. My orange wool hunter's cap was snatched from my head to dangle on a limb, branches caught my hair, twigs snagged in my coat collar or my shirt and held me immobile. Thorns ripped holes all through my orange plastic hunter's vest. Leaves and limbs slapped my face, threatening my eyes. Beggar's-lice and stickers found convenient resting places on my legs and rear end. I heard strange noises all around me. And just as I really felt spooked, Wiener would bound up, chasing a squirrel or a rabbit right to me. But I never saw one deer!

When I emerged from the woods, Dennis and Virgil would report seeing the same three animals: an old doe, a young doe, and a very young "button" buck whose horns had barely erupted. They would not even attempt to kill such creatures and were deeply disappointed at the lack of full-grown bucks.

The last day we hunted, Dennis decided to have me run the ridge in the opposite direction going as far as the end of our property beyond the creek. Several years before he had shot a ten-point buck there whose near perfect antlers—a trophy we

45

have hanging on our wall—won him permanent membership in the Missouri Big Bucks Club. I had never approached his stand in this spot by descent from the top of the ridge and I was wary.

I ducked under the cattle fence that crosses the ridge and the barbed wire tore a two-inch rip in my pants leg before I'd even gotten started. I waited by the old family cemetery and when the time came, skirted the cedar thicket on the ridge without mishap, only to slide down the other side on my rear end, so that my pants were now wet as well as ripped. I used my walking stick to get me past the food patch—a small five-acre field where we grow grain to feed the deer and other wild animals—partway down the ridge, through a deep ravine, past some jutting rocks and right up to the tree where Dennis had his deer stand. My orders were to wait there for ten minutes, then to walk fifty yards straight ahead, then to turn right and walk into the upper end of the hay field. I tried it . . . four times. And each time I ended up right in front of the deer stand again.

I knew that it was nearly noon; Kevin would be returning home from kindergarten to an empty house. Wiener appeared —and disappeared again. I dropped to the ground beneath the deer stand and lay there, exhausted. Clouds were moving in overhead. The sun had vanished and the tops of the trees began to sway in the brisk, quickening breeze. When a few drops of rain fell on me I crawled into a cedar thicket trying to figure out which way to move. I could hear cars traveling along the highway past the upper end of the valley. I even heard

Dennis drive his pickup in my direction. I felt foolish and frustrated. I knew where I was but I could not figure out which way to walk to get off the ridge.

Finally, when the rain subsided, I started backtracking past the jutting rocks and down the ravine, which would, eventually, have to bring me into the hay field behind the hog pens. I had nearly made it when Dennis came into view, hunched over the steering wheel of his pickup, bearing down on me at full speed. I had to scream to flag him down.

His first words were, "Where in hell have you been for the past two hours? I was just off to get Dad and some men to start looking for you along the highway."

I would have had to cross several acres of wooded ground on someone else's property to reach the highway. I wasn't as dumb as all that.

"I know better than to go way up there," I answered.

"You must have gone pretty far. I even honked the horn and you didn't answer. Boy, I can't figure out how you could have gotten lost when all you had to do is just come down the ridge."

I had heard wind, rain, squirrels, cars, Wiener barking, the pickup engine, but no horn.

"I was not lost. I knew exactly where I was all the time. It was your dumb directions that got me messed up."

"After all this time, I figured you could find your way off the ridge."

"And I did, didn't I?" I snapped back.

At about this point I realized that Wiener knew the way Dennis always left the ridge; I should have followed him.

But Dennis was furious that I'd talked back to him—or maybe it was really the aftermath of having been frightened that I had fallen and been injured when I hadn't shown up. He jerked the truck into gear and sped off, leaving ruts six inches deep in the dirt road.

"Next year," I declared, "I'm leaving the country during deer season."

Later in the week Virgil did manage to shoot a six-point buck but after it was divided among five families we had little more than a taste of sausage and steaks.

Dennis, Jack, and Kevin consider deer steaks a favorite food. I think venison *can* be as good as beef. But it took me a long while to learn how to prepare it. I have canned venison, which many people consider delectable. I don't, at least not any I have canned. I have cooked it in a slow cooker but that leaves stringy meat and a "wild" taste, which some people object to. I find the best method is to mix to taste several cups of flour, ground garlic, pepper, basil, and a little salt; sprinkle the meat with Worcestershire sauce; then pound the flour and spice mixture into the meat with a meat cleaver. Pound both sides at least twice until the meat is thin and flattened. Brown in a skillet with a few tablespoons of vegetable oil. (I use an electric skillet set at 380 degrees.) When the meat is brown, cover with water and simmer for about two hours, being careful to turn the meat and to make sure the liquid doesn't cook off. This makes a spicy, tender dish with its own fairly thick gravy.

We make most of our venison into sausage, adding some pork or beef for consistency rather than taste. The men who do the butchering of the hogs and the deer do most of the work, but the women help by cutting up the meat, packaging it to freeze, or stuffing the meat into the casings.

Venison, like other wild meat, is very lean. Nutritionally, it is low in cholesterol and extremely healthy but it needs a binder to make good sausage. We take deer sausage along when we work in the fields as energy food. It stays fresh for days without refrigeration and is handy to have when there is no time even to sit down to eat a sandwich.

One Saturday of the season, Dennis's family—his parents, aunt and uncle, a cousin, and his two sisters, Alice and Karen, and their husbands and children—will gather for a meal at our home. My boys and their cousins—Karen's two daughters and Alice's two boys and a girl—match in ages, and they always have a lot of catching up to do. Karen and I get the evening meal ready while Alice will go out in the woods with the men. Alice *likes* to hunt!

# Chapter Five

We have a Thanksgiving tradition on the farm. Thanksgiving week we begin to chop firewood. This takes quite a bit of work since this is the wood we depend on to heat our home.

The first few years after we built our house we had no chance to cut wood ahead of time so it could be stored to give it a chance to dry out. We ended up burning green wood. Often such wood has laid on the ground under several inches of snow. It produces a small, smoky fire. After the flood of 1986, my stepfather chopped up the trees on his property that had been trashed when the levee broke and delivered the wood to us. But this wood was so wet that when we burned it, water poured out of it. We let a pile stand for the summer, hoping it would dry out. It shrank to half its original size, but when we put it in the stove, instead of burning it just sizzled. When finally it did burn, instead of leaving a residue of ashes it melted into large flakes that filled the stove so that we were dumping ashes every day instead of once every two weeks. We learned our lesson, and now we try to cut the wood we will need one winter ahead and store it in a dry place until it is wanted.

Whether we're using the chain saw or ax, loading or unloading, or stacking for storage, these chores are best performed methodically. All that walking, stooping, lifting, bending, and swinging will eventually reach into every muscle in your body, giving you the best workout you've had for months. On a warm sunny day when you're alone in the woods, it can be an inspirational experience. It can also be tiring and stupefyingly monotonous. That's when it gets dangerous. We're aware that it's risky to work with equipment that's not sharp or running properly, or in poor light, or when your mind is wandering. Luckily, we have had few serious accidents. But a few years ago, my father-in-law had a bad one. He was cutting wood alone. Somehow the chain saw slipped. He needed two dozen stitches in his knee.

That is not to say that minor injuries are uncommon. Right now I am nursing a scraped hand and a bruised right shin as a result of carelessness while handling chunks of wood. And Dennis has a dead fingernail falling off that he banged somehow several weeks ago.

When we prepare firewood we try to select a dead or dying tree, preferably one that has fallen. If the tree must be cut down, the cut is angled so the direction of fall can be predicted. Dennis uses the chain saw to cut the trunk and larger limbs into logs. Then the short chunks have to be stood on end and split with an ax into halves, quarters, and eighths, which is work I can help with.

Many local people use log splitters. These funny-looking contraptions travel on two wheels. Some hook up to the power take-off of a tractor; others come with their own gasoline-operated motors. Like most farm machinery, each manufacturer's design varies; all have an attachment that splits a log cleanly, as if with the blade of a knife. But we still use the old-fashioned method: ax, arms, and back muscles.

When we can, we work as a team: Virgil uses the chain saw, Dennis splits the logs with a sixteen-pound maul, and Jack, Kevin, and I load the pickup and unload it to make a woodpile by the house or carry wood into the woodshed. This turns out to be a real family project for us. Just as well. Cutting firewood is a job that is never really done. Even when we have enough wood stored for the next winter, we cut and store for the winter after that whenever we have a bright sunny day in November, December, January, February, or March.

. . .

December is always an exceptionally busy month for me. Harvest is ended and we must prepare for the next year, the end-of-year bills must be paid, and we have to get ready for Christmas. I like to make most of the gifts we give to others so each year I have to hurry to finish crocheting scarves, hats, shawls, and afghans. Before we had quite so many nieces and nephews, I sometimes sewed each one an identical shirt. Now the cousins exchange names, and gifts, with one another. I always hope my boys will draw a girl cousin's name; it is the only time I get to sew something pretty.

My mother's fourteen brothers and sisters, my stepfather's family, Dennis's family, and our many, many friends make our Christmas-card list a long one. We are lucky if we see most of them once or twice a year, so I like to add pictures and a note to our cards. It can take several days to complete the list and get them all addressed.

The buying and wrapping of gifts has to be done, and in secret. Santa is still a visitor to our house. There are get-togethers with friends, family, and neighbors. And there is baking. Shortly after we were married, Dennis asked me to learn to bake bread, and after some trial and error, I mastered it. Most of our friends give cookies as gifts but I bake bread and stollens.

December is busy; in addition to holiday preparations, there is the regular work to be done. My schedule for a typical December reads:

Make firewood—10 hours
Clean house—15 hours
Sell hogs,
other farm work—32 hours
Christmas cards,
making gifts—16 hours
Shopping—11 hours
Baking—20 hours

Plus six get-togethers with family and friends or at 4H and school events and two days looking for and decorating the tree

One year, I spent the best part of the Sunday before Christmas looking for a good live tree for Dennis to chop down for us even though I was still working on my Christmas cards and I hadn't finished my shopping. We also had a visit from my Aunt Elizabeth early in the day and then went to pay a visit to Jack's paternal grandparents that evening. The next day I finished my shopping and cards but I had to go to the dentist on the twenty-second so I ended up decorating the tree with a very numb mouth. The next day I began baking: eight loaves of bread, four stollen, two dozen cupcakes, and four batches of cookies.

Then Kevin announced that he wanted to make a cookie house. We started on this project the evening of the twenty-third. We made the walls and floor from vanilla wafers, the roof from sugar cookies, the shrubs and trim from jelly beans and sugar candies. The fun was in splattering the icing to cover the aluminum-foil-and-cardboard base and to stick the cookies together. But getting the chimney—a stack of round peppermint candies—to stand up defeated us. Even using candy canes as braces didn't work. Still, Kevin didn't seem to mind this flaw, and he and his cousins demolished it cheerfully on Christmas Day.

Christmas Eve I baked more stollen and some cookies. My parents came to visit, and later that evening the boys and I attended Christmas Eve Vigil Mass.

Unlike the women who are my friends, few of the men I know attend church scrupulously. Dennis belongs to the United Church of Christ and does not come to Mass with us. When we married I assumed the responsibility of raising the boys in the Catholic Church, and I do my best to fulfill this duty.

Coming home, we passed the 7-Eleven and IGA stores, still open, but had no reason to stop. But as soon as we reached our house, Dennis greeted us with news of a crisis.

The farrowing room is almost empty at this time of year. Most of the sows are between seasons. But there are always a few who farrow in the off-season. One, who had just given birth to eight little piglets, had suddenly had a seizure and died. Sometimes this happens; pigs too have birth complications. Ordinarily, Dennis would place the newborn piglets with another sow that had recently farrowed. But this Christmas Eve there weren't any. The orphans seemed healthy, but if they were to survive we would have to feed them by hand.

They were still in the pen where they had been born, under a heat lamp. They would remain there. Jack went to add a couple more heat lamps to try to simulate their mother's body heat. I drove back to town to pick up a case of canned milk. But by the time I got there, all those invitingly lit-up stores had closed.

The only milk I had in the house was two gallons of two-percent low-fat milk, and even homogenized milk does not contain enough fat and nutrients to sustain a baby pig for very long. So as soon as I reached the kitchen again I got out our biggest pots and pans, into which Dennis poured the milk, adding boxes of brown sugar to it. He would have to feed those pigs this mixture every three hours throughout that night, the next day, and several days thereafter. And I have complained that Dennis has no patience! But I learned better that Christmas Eve as I watched him take each little pig into his big hands and gently and repeatedly nudge its snout into the milk-and-sugar mixture until the piglet learned to sip the milk. Sometimes it took five or ten minutes for the pig to get the idea. And by the time the next feeding came around, three hours later, he had forgotten, and Dennis had to coax him to use his tongue to sip at the milk all over again. So that night, while the boys slept and I played Santa, dragging out the gifts I had hidden away, Dennis played mother to eight little pigs.

On Christmas Day we made a fast trip to Jefferson City to have dinner with Karen's family and an even faster trip back, holding our breath, hoping the piglets were all right.

A few days later another off-season sow had a litter. Hers

were stillborn, but she had plenty of milk, so Dennis placed the orphaned piglets with her and she nursed them happily. Not one of them died before they reached a weight of two hundred and fifty pounds apiece; then we sold them at the local market.

And now I always keep several cans of Pet evaporated Vitamin D–added milk in the kitchen cabinet in case of another emergency.

After Christmas, Dennis and I spent several days in the woods selecting trees to be cut for timber (his purpose) or to be saved (mine). The hillsides on our farmland shelter acres of huge trees; most are more than a century old with trunks over three feet in diameter, their bare-branched crowns looming several stories high. But unfortunately, many were diseased and dying. We had to decide whether to clear out the bad ones while we could still get a fair price for the timber or face the prospect of an entire wood of dead trees in a few years. Even so I was determined to keep as many as I could. I tramped the woods marking with twine the trunks of the ones to be spared by the woodcutters.

Dennis said, "I should have said to mark the trees you wanted them to cut down. It wouldn't have taken near the time or twine because you wouldn't have marked a damned one."

I had no comeback; he was right.

# Chapter Six

The depth of winter can be a sad time of year. One bleak day at the end of January, Dennis's grandmother died. She was ninety-five and had been in a nursing home as she was unable to speak or help herself in any way. She didn't even recognize Virgil, her only child. She had been in a vegetative state for years. In fact, she had laid in the fetal position for so long her legs were frozen against her chest, so she had been embalmed that way and that was the way she was laid to rest. It upset me to think of that poor woman going through eternity with her legs in the air. I couldn't seem to tear my eyes from the sight, but Dennis, who had been close to his grandmother, took one look and didn't go near the casket again. When Jack and his cousins ventured near, they pretended to study the flowers at the foot of the bier, but Kevin got right up there with his nose in the lace that trimmed the pillow. I held my breath, knowing he had never seen a dead person before. He put his hands on his hips finally, and declared, "Yup, she's dead. They'll bury her tomorrow."

They did. Jack, as well as Dennis, was a pallbearer.

A week later I learned that the man who was my biological father had died of a heart attack. The man I call Daddy is the man my mother married when I was four years old. My biological father, Roy, deserted my mother and their three children, of whom I was the youngest, when I was fourteen months old. He disappeared without a trace, leaving us destitute. For nine years we didn't know if he was alive. But when I got older I kept up a regular correspondence with my paternal grandparents. A few weeks before I graduated from high school I received my first communication from my father. He sent me a check for twenty-five dollars and asked me to call him, collect, at a number he had written down. I did. I telephoned from my high school; I didn't want to get into a discussion about this at home. It was odd speaking to him when I had no memory of him. Four years later, Grandma died and Granddad told me he'd gone to live with Roy. Then Granddad died and I lost touch again.

I had never met my father to remember him. Now I was thirty-eight, and my last chance to see his face would be as he lay in his coffin.

What little I had been told about Roy was that he had been a glider pilot in World War II, served in the Air Force for more than a decade, and then received a dishonorable discharge around the time that I was born. My brother, who is eight years older than me, joined the Air Force and traveled extensively; we had had little contact. It was assumed that he would attend the funeral. However, my mother and my older sister decided not to go and my mother forbade me to attend it. But this was a decision I had to make for myself. Unfortunately, it was not a decision I could ponder at my leisure.

Dennis, the boys, and I were nursing our annual winter colds, and Dennis and I had been trying to get the book work done so the accountant could prepare our income taxes—farmers must pay taxes on March 1. And the long auger in the finishing building had coughed its last; we could no longer put fourteen tons of feed into the large bin to supply the whole building for several days. So to feed the hundreds of hogs housed there, we were running the auger attached to the feed grinder through the window nearest each feeder and filling each one directly. As different size hogs require different feed, the fact that a full bin holds nearly five days' worth didn't actually help to reduce the burden of work. We ended up doing this chore on a daily basis. And it was tricky. Dennis drove the tractor pulling the feed grinder so I, on foot, had to reach up to try to guide the auger; one miscalculation and the auger could ram right through walls or the roof.

Several inches of snow had fallen a few days before. The temperature had then dropped until it reached subzero, causing the water pipes in the hog houses to freeze. We had to spend six hours moving different size hogs into pens in the four rooms of the finishing building so that the heat of their bodies would raise the temperature inside uniformly and thaw the water pipes. Virgil drove into town to purchase Styrofoam sheets of insulation that he and Dennis packed into open or broken windows. I just watched because I had no insulated work boots, and outdoors, in that temperature, my toes would have frozen.

The next day we began to fill the feeders in the finishing building *again*. It was so cold! My poor feet nearly broke off at the ankles. But the water pipes in the finishing building were thawing; steam was even beginning to rise around the roof just from the hogs' body heat. So long as the doors and windows remained closed, the temperature inside would now stay at about fifty degrees no matter how cold it was outdoors.

Now the feed grinder became the problem. It is an odd-looking contraption on two wide wheels with a funnel-shaped pot about ten feet tall and six feet in diameter at the largest point. It can hold approximately two and one-half tons of feed. On its left side is the twenty-foot folding auger that carries the ground feed from pot to bin. Input hoppers are located on the right and in back of the pot. A pulley wheel and the hookup to the tractor power take-off (PTO) are located in front. The entire working of this gigantic contraption depends upon one two-inch-long, one-quarter-of-an-inch-thick pin that connects the pulley wheel to the grinder. A few days earlier, too much corn had been spilled into the feed grinder so the plates located between the PTO and pulley wheel could not crush fast enough. They clogged and the pressure broke the pin. The local store was out of pins; we had been waiting for replacements. Virgil picked some up when he went for the Styrofoam but we didn't have time to try to repair the feed grinder until the next day.

Lying on the snow-covered ground, Dennis squeezed his six-foot-two, two-hundred-and-sixty-pound body into the one-foot space behind the front wheel of the grinder and tried to push the pin through the tiny hole in the pulley while I slowly turned the wheel to line up the holes. But on such a cold day both plastic and metal are so brittle that they tend to snap. This happened to us several times. Each time, Dennis climbed up the ladder, unlatched the trapdoor on top, and emptied some feed out by hand to make more room inside the grinder. It took until midafternoon—and five broken pins—before we got going. A long cold day.

Dennis drove the tractor while I guided the auger from outside the building, with Virgil doing the same thing from within, to fill the ten three-by-six-by-five-foot feeders to the brim with a week's worth of feed.

At about seven o'clock in the evening, while Dennis was still in the tractor grinding feed, Virgil and I went to sit in the pickup to warm up and we began to talk about whether or not I should go to Texas for my father's funeral. The quandary had been preying on my mind all day, and when Virgil told me that in my place he'd go, I was finally able to decide.

I ran to the house, called my parish priest and my friend JoAnne to arrange coverage for the next week's CCD classes, spoke to another friend, Paula, packed our suitcases, and was ready to go by the time Dennis came in for supper. He was neither surprised nor displeased. He had often urged me to stand up for myself against my mother and sister, and I think he agreed with Virgil: I owed it to myself to go.

We left Wednesday morning as soon as the boys were on the school bus. The further south we traveled on Interstate 44, the warmer the weather became. By the time we reached the Missouri-Oklahoma border, our heavy coats came off. Oklahoma viewed from the Will Rogers Turnpike, seemed all brown terrain and Wal-Marts every few miles. We passed several motels advertising themselves as "American Owned" or "American Operated"; we couldn't figure in comparison to what, or why anyone would care. We stayed in a motel that night, and the next day we reached De Leon, Texas, a town a hundred miles or so southwest of Dallas.

We stopped at the first pay phone to call the funeral home and get directions. Soon after we reached the motel where reservations had been made for us, my brother and his wife arrived. We four had dinner and then went to the funeral home.

The funeral home occupied the first story of a white frame

house. The interior was carpeted in a soft periwinkle blue with attractive draperies and furniture. Several large bouquets already stood at the ends of the casket, which was made of a blue-gray metal. Inside it lay an old man. Just an old man whose remaining hair was gray and white, with glasses perched at an odd angle on his nose. He looked a lot like my brother and like his brother, my Uncle Mac.

I looked at him for a long time. I felt numb. I knew that when the time came for me to look down at the body of my stepfather—the man who used to take me into town almost every Saturday to buy me a soda, who taught me to stop a runaway sled on a hill, who wept with me at the death of my first husband and applauded my decision to marry Dennis— then I would be in tears. Now I hardly felt anything. My brother had some memories of Roy; I had none. I wept but not out of grief. My tears were for the sorrow of never having had a chance to know the person named Roy.

Later, when my brother and his wife and Dennis and I went over the funeral plans and expenses with the director of the home, I saw Roy's discharge papers and learned that contrary to my previous information, he had voluntarily resigned from the Air Force, no one knows why.

Later still, back at the motel, I met the aunt and uncle about whom I had always been curious, who were there along with another aunt and uncle whom Dennis and I had visited on our honeymoon.

The funeral was to be held in what was said to be the oldest Baptist church in that area. It was a brick building. The interior looked suspiciously like a Missouri community dance hall: The floor was waxed hardwood, the windows reached from floor to ceiling, there was a short row of benches and a low stage at one end where the casket rested next to a piano, microphone, and lectern. There was a colorful mural on the back wall. I am a Catholic girl, used to stone vaulted churches, carved and gilded statues, stained-glass windows. I was surprised.

The service was brief; the eulogy moving. At the graveside, the Legionnaires took over. Roy was buried with full military honors. At the cemetery, Uncle Mac and Uncle Oscar pointed out several headstones belonging to my ancestors. We drove out to the old homeplace, a deserted stone cottage on the prairie. Then there was a Texas-style gathering of family with plenty of food. The next day we paid some family visits, but by Sunday evening Dennis was anxious to start the long drive back to our farm.

This time we headed east through northern Louisiana and then due north through Arkansas. It was dusk or dark as we went through Louisiana, but I remember the road was built high through the wetland and that it was so warm we kept the windows cracked despite the rain. It was strange to hear frogs singing in February; the odors and mist reminded me of a Missouri spring.

After a night in a small town south of Little Rock we reached the interstate and entered that city right at the noon rush hour. For the first time in eight years I heard my husband pray. He said, "Lord, help us get through this son of a bitch."

Dennis was indignant when I laughed. But of course we made it, as he is a very good driver even though he has to depend on me to tell him what color the traffic light is and we are unaccustomed to heavy city traffic.

I had time on our return journey to be glad that we had gone. I had found the other half of my family, people who had known my elusive father in the days before the pain and sorrow had begun.

I think most of his descendants have some of Roy's wanderlust—and some fear that unless we resist it, it could destroy us, too. For myself, I know that I could not exist without my husband and sons. My freedom lies in our life together here on the farm.

# Chapter Seven

We refrain from making pets of our pigs. Jack discovered the reason for this the hard way. The first year he raised barrows (castrated, fully grown hogs) to show and sell at the county fair, he tamed his two to the point that they would follow him everywhere, let him use their bodies for a pillow, and even drink water from a cup he held out to them. When he sold his first pet for $450 and the buyer began to load him onto his truck, it suddenly dawned on Jack that his pet was going to be someone else's bacon, ham, and pork chops. It was more than he could bear.

Nevertheless, the boys and I have bestowed names on many of our sows and boars because they are around much longer; they often die in the pens of old age. And hogs do have distinctive marks and mannerisms and, under the right conditions, will become as tame as any other domestic animal. We had a purebred Large White boar we called Big Daddy who weighed over seven hundred pounds and had tusks six inches long. He was as strong as a full-grown bull but so gentle we could pet him. I think he cared for his sows as individuals too, because I often saw him nuzzle them as they ate instead of just swiping them out of the way with his massive head as most boars will. And if his sows began to squeal and fight he would jump between them as if to say, "Now, girls, can't you try to get along?"

Big Momma was a spotted sow who stood so tall (on her four feet) that her back scraped the top bars of the crate in the farrow house. We also had to adjust the length of the pen to give her more room. She weighed between seven and eight hundred pounds and her eyes were blue. She had no special breeding but was so tame she begged to be petted. She repeatedly produced litters of at least ten piglets that were the largest and healthiest in the building.

Tusses, Scritchy, and George were purebred Hampshire boars, each of whom earned a nickname. Tusses got his tusks caught trying to get through the fence and had to be cut out; Scritchy demanded to have his ears scratched and would glare fiercely and butt your leg or side until you complied; George was just plain stubborn. He always took his own sweet time moving and had wrinkles all over his face like a Shar-Pei puppy.

Piglets are cute and cuddly for the first six weeks, but they don't really like to be handled. Most will squeal themselves silly at a mere touch. But as personable and cute as an individual pig may be, dealing with hogs inevitably involves the performance of a lot of unpleasant jobs.

One chore I really hate is castrating, but it is necessary. Boar meat emits a strong, noxious odor and is often unpalatable as well so a boar does not fetch the price of a lean, fully grown two-hundred-and-fifty-pound castrated hog. We castrate most of our pigs when they weight thirty pounds or less. I can catch them easily then, and Dennis and Virgil can perform the surgery without any trouble. The bigger the boar, the harder he is to handle, and fully grown boars have sharp, pointed tusks. There is a technique to grabbing a full-grown boar, throwing him down, and overpowering him, but it takes brute strength equal to that of a hog.

We have hosted parties where hog nuts or mountain oysters are the main dish. The texture and taste are comparable to chicken livers but I don't care for the aftertaste, which is bitter: the larger the hog, the more bitter. Most of the men we know like them and even request them; most of our women friends agree with me that we can live without this dish.

Another disagreeable task that often falls to me is caring for sick piglets. Trauma during or immediately after birth gives piglets scours: diarrhea plus scaly skin patches. Pneumonia is also a serious problem for the young unweaned pig so we guard the litters against drafts and wetting. Newborn piglets must be kept in a room with a constant temperature of seventy degrees. Even with all these precautions they sometimes contract viral pneumonia. Piglets with pneumonia are not difficult to handle, but when very ill their bodies become cold, still, and lifeless except for their laborious gasps for air. When they are this ill you know ahead of time that they will die, and it seems kinder to kill them to end their suffering. Often we lose a whole litter no matter what we do to try to save them.

A hog's internal structure is similar to that of a human being, and like human beings a percentage are born with congenital deformities: missing limbs or eyes, heart malformations, tumors, and the rest. We once had a piglet with two faces: one skull, two ears, three eyes, two snouts. Usually these animals are born dead or die in a few hours. The most pitiful birth defect is the lack of a lower bowel and anus; the poor animal has no outlet for wastes and gets wider and wider until it suffocates or explodes internally. Surgery is sometimes attempted, but rarely works. A good number of hogs have abdominal ruptures but unless the rupture breaks through the skin a hog will survive. (We usually hold these animals back to butcher ourselves as they fetch a reduced price at market, although none of the meat commonly eaten is affected.)

Pigs are also vulnerable to an intestinal virus that strikes when they are weaned and their feed changed. More than once we have lost as many as two dozen in the course of a day.

This represents half a truckload of hogs had they reached maturity, a loss of at least five hundred dollars. We do our best with vitamins and antibiotics for a particularly puny-looking animal, but when we lose them in quantity like this it's a severe blow.

Even older hogs are highly susceptible to many diseases, some of which are carried by man. We keep a bowl of disinfectant near the door to the farrow house in which to rinse the soles of our shoes to kill any bacteria or the larvae of intestinal worms that we may have picked up elsewhere. Salespeople and feed deliverers who have tracked around other farms must use this bowl. The vulnerability of our hogs explains why we are not very welcoming to strangers and discourage visitors to their quarters.

One of the worst jobs is having to clean the dead animals from the pens. Hogs are cannibals, and a sick or injured pig will be bitten and chewed on by the others until it dies. Then they will eat the carcass. It has to be dragged away from them, taken out, and buried in a plot we have set aside for this purpose. I do not like this job.

But of all the jobs a hog farmer does, I think the one I hate the most is sorting, and it's a job I am most often drafted for.

Hogs have to be sorted for many reasons. Sows must be segregated according to breeding and farrowing schedules. (Several boars can be kept together provided they are used to one another so we generally buy our boars in small groups and keep the group together for a lifetime. Strange boars will fight until one kills the other.) Sometimes an animal that is ill or injured can be sorted out from the rest—who would otherwise kill it—and kept apart until it recovers. Occasionally, we run feed tests for various companies and then hogs must be sorted and kept in separate pens so the test results will be accurate. But primarily we sort our hogs by size, to sell.

When we returned from Roy's funeral of course the feed grinder still needed to be rebuilt and the year-end book work and tax preparation awaited me, but what was far worse was being confronted with a large number of hogs waiting to be sorted before they could be sold.

They had to wait a few days longer; our business had enlarged so that Dennis needed more and faster communication of information than that provided by our local radio station or a weekly printed commodity report. So we decided to rent a KU-Band computer screen and satellite dish. The equipment was shipped to us via UPS. It came in three boxes. One was delivered when we were not home and was left out in the rain. The others were set in front of the door to bake in the sun. But as soon as we received those boxes, Dennis, who couldn't wait to get started, spread all the pieces out in the living room and ordered us to not so much as breathe on them while he figured out how to assemble them.

The next morning I went to town to do some errands. When I stopped at the mailbox on my way back I could hear the electric drill. Dennis was putting a hole in the roof for the satellite dish cord. He finally got all the pieces—nuts, screws, bolts, wires, frame—together but they didn't fit the satellite dish and the computer wouldn't work. We had to call the company, which promised to send new components. The frame fit the new pieces, and the disc seemed right but the new visual projection piece looked nothing like the diagram. I dug into the first box and there, sure enough, was a piece like the one in the picture. We attached it to the satellite dish, the cord to the screen, and we were in business. Fricke Farms had entered the computer age.

Dennis hired someone to repair the grinder. But though I tried a couple of times, ultimately there was no way I could manage to evade hog sorting.

So as soon as the boys boarded the school bus I donned my raunchy clothes and rubber knee boots, put on a knit cap, and trudged down to the finishing building where Dennis was waiting to spend two or three hours separating hogs until Kevin returned home at noon from kindergarten.

Sows and boars—mature animals kept for breeding—are generally slow, lumbering individuals accustomed to being moved from one pen to another or in and out of buildings. They seldom put up opposition. They're usually gentle and go where directed without a fuss. But a three-hundred-pound hog is very strong and it's best not to forget this. The general rule is to stay behind the animal you want to move, but sometimes this is not possible. When we shift the sows in the farrowing building I must often act as a gate to direct a sow into the right crate. If she doesn't want to enter, I don't argue with her. I let her wander around the entire room again rather than risk being shoved onto one of the steel crates.

The gilts and barrows—young hogs kept not for breeding

but to be sold at market as soon as they reach the proper weight—become wild when they are moved. Probably because they haven't had enough experience to know what to do and they're frightened. They may weigh less than the hogs we keep but they are big enough to be dangerous if you're in their way. And even worse than the risk of direct injury is the air pollution that results from moving hogs indoors. There is always a noxious odor from wastes in the hog buildings that is most noticeable in the humid days of summer. And whenever gilts and barrows start running to avoid being moved, they kick up dried feed and wastes, and the air inside the building becomes thick with dust.

Many hog farmers wear masks for this task, but we have found masks to be an uncomfortable nuisance. We do without or tie a scarf over our faces. But after several hours at a stretch in the finishing building I suffer from coughs and chest pains. And Dennis's sinuses have been almost completely destroyed by constant exposure to this dust and to gases. Now he can't smell anything, not even my favorite perfume.

Dennis can look at a hog and tell how close it is to two hundred and fifty pounds—selling weight. He can also judge a sow's farrowing date within a few days. This is still beyond me. So generally I guard the gate to the pen while Dennis walks among the hogs and judges them by size, length, height, and width. Hogs seem to prefer to travel in a group so to get the hog under consideration to leave the pen it's easiest to let some others out with him. They will only try to storm the gate if you don't. On the other hand, if you want the whole dozen they will ignore an open gate and refuse to budge.

Dennis will walk the hogs to an open area near the loading chute where he studies each one to decide which to sell and which to return to the pen. This sounds simple but these are young, frisky hogs, and by the time they have been run around the pen a time or two, walked out to the loading area, and herded back, they are running wild.

This is not a job we perform in silence. The sequence sounds something like this:

"Out!"

"In!"

"Not *that* one!"

"Which one?"

"The one between those other two! Here they come! Stop 'em! Open the gate, open the gate! NO! Close it!"

"Damn it, Dennis, make up your mind!"

"Will you shut the damned gate before you let the whole pen out!"

And so on.

Our farrowing house is designed so that as the hogs get bigger they are moved into larger pens, the largest being the pens closest to the loading chute. Our pens have gates on either side of the feeder. The concrete floors beneath the feeder and feeding area slope so as to drain down to a slatted area over a pit. Feed and animal wastes wash through the slats to the pit and thence into the nearby lagoon. But this means that not only do I chase hogs in circles, I have to chase them up and down on an incline on a floor littered with spilled food and feces.

The trick to getting a stubborn hog out of a pen is to get it running back and forth so that it becomes confused and will duck out of any exit it sees. The drawback to this method is that the person chasing the hog has to run at top speed while making the same abrupt spins and turns as the hog. It's filthy and exhausting work.

I carry a piece of plastic hose with which to whop the hogs when necessary. This is not cruel: Hogs have incredibly tough hides. There is good reason for pigskin being the material of choice for luggage, shoes, and the like. Swatting a hog with a piece of plastic to get his attention hurts me a lot more than it hurts him. Swinging the plastic tubing tires my elbow, and the impact jars my wrist. All the pig feels is a little vibration. Besides, usually all that is needed is a touch or even the mere wave of an arm, or the hose, in front of a pig. It's just as well, because, hard as it would be to bruise an animal or cut its skin, if this did happen, he would be unsaleable.

When the hogs are sorted and ready to be sold we load them onto one of our thirty-thousand-pound-capacity farm trucks, a two-ton with a hoist bed. We try to do this early in the morning, before sunrise, for a couple of reasons. In summertime this avoids overheating the hogs and the earlier we wake them the easier they seem to be to load. Also, the local buyers close down at noon.

Two people can manage the task of loading. But it's a lot

easier with three. Then one of us guards the gate while Dennis sorts out three or four hogs at a time and ushers them to the loading chute while the third person stands guard at the end of the chute to block off the tailgate of the truck so that the hogs on board can't come back down the chute. Dennis usually drives them to the local market, where he has to unload them alone but Virgil sometimes takes them to Central Missouri Livestock, about fifty miles away, where help with the unloading is provided.

Sometimes hogs walk right up the chute into the truck. More often, there is at least one stubborn one who tries to turn back or worse, gets halfway up and refuses to go any further. Then Dennis uses a handheld battery-operated shocker to get the animal moving; when applied to a hog's rump it does its job well. Occasionally a hog tries to jump over the side of the loading chute; few succeed but those are even harder to corner and reload.

We finally sorted the largest hogs out of four pens but other farm chores, the weather, low prices, and our persistent colds delayed the loading for nearly a week. Finally the evening came when Dennis said, "We gotta sell 'em, Mim. We'll never be able to get them on the truck if we don't load them now."

I sighed in resignation. I knew he was right.

The larger a hog is, the more fat it has, and today the market demands lean animals. If we didn't get the hogs to market they would get too large and money would be docked from the sales price for the excess over two hundred and fifty pounds. So although the weather report predicted that the next day would be cold and blustery, with rapidly falling temperatures, we had no choice. We had to get our hogs to market.

Six-thirty in the morning came far too soon. My consolation was that in summer we would have had to start at five. And we had only one load today. Sometimes we load two or three trucks in a morning.

Dennis was up and out the door before I crawled out of bed to dress and wake Kevin. Jack was already up. I made breakfast and packed Jack's lunch. I took Kevin with me to the hog buildings so that the two boys would not distract one another and both miss the school bus.

Though sunny, it was brisk. I was wearing insulated socks, boots, jeans, a knit shirt, two sweatshirts (one with a hood) and a pile-lined jacket plus a woolen cap and gloves but I still felt the wind whip through me.

Dennis already had the truck backed up to the loading chute and was waiting for me to take my place beside the chute. Since it was going to be just the two of us, he would prop the gate open and run the hogs from the pen by himself.

The first group walked out of the pen, through the loading area, up the chute, and onto the truck as if they had been rehearsed. So did the second bunch. The third group included a stubborn hog that stopped in the middle of the chute and then, when prodded into the truck, refused to stay there. By now Kevin had four minutes to get to the place where the school bus would stop for him. Dennis asked me to wait to help him load the next bunch, but how could I? Instead, I got Kevin started toward the road on a shortcut across the hay fields; I

78

hoped Jack would stall the bus driver and, if she saw Kevin coming, she would wait.

When I got back to the chute, Dennis had four hogs waiting and was swearing because the gate was up and I wasn't there to open it. When I tried to lift the board to open the gate, the movement startled the hogs and they tried to bolt down the chute backwards—along with the recalcitrant one from the previous bunch. Finally, using pipe, prods, and a plywood board as a shield, we managed to get the five of them onto the truck. And then our nemesis bolted again and ran down the chute beneath the four newcomers and between Dennis's legs, back into the pen.

The next group had at least six hogs in it—I was losing the ability to count. And then two mounted the chute while four turned around at the bottom and just stood there. I had to whop them with the pipe and then take a swipe at one of the pigs at the top of the chute who was trying to back down again.

Dennis brought out six more but now *two* were trying to get off the truck. One pig was descending the chute while three were on the way up. They got stuck in the middle.

I'm not a good climber; I have short arms and legs. But I did my best to crawl up the side of the chute full of squirming, squealing hogs and bring them to order with my piece of plastic pipe. I was too slow. One of them managed to back down the chute and the rest followed, knocking Dennis, his plywood shield and shocker to the ground. Now I really got cussed.

"If you could just do one thing right!"

I thought of all I did do right and I yelled back, "The trouble with you, Fricke, is that you expect the impossible. I'm just not tall enough or strong enough to climb over these hogs and beat them up the way you would!"

"A person might think you'd have learned how by now!" he yelled back.

"If I could, you can be sure I would have, long since!"

Still he grumbled, "I just can't be in two places at once."

"And I can't hold open the truck's tailgate, beat the hogs at the foot of the chute, and climb up in the middle all at one time either!"

He stomped back to the pen and I gasped in disbelief. I had actually managed to have the last word.

But I had no time to dwell on this victory. Dennis was back with more hogs.

Another stubborn bunch: I could no longer count them. Too many hogs were squealing and fighting on and off the truck. They were louder than Dennis and me now.

We finally got down to six, and Dennis asked me to come help him to run them out of the pen. We had a ten-minute chase before one leaped out of the pen and down the aisle to the loading area, followed by three friends. That left two and we decided we'd come back for them later. But when we reached the truck, the board over the tailgate had been knocked down, and a hog was about to walk back down the chute. I managed to get to the tailgate of the truck in time to whack him on the snout as he was about to step down and turn him, but then two others ran past Dennis toward the pen. . . .

It took us another quarter of an hour to finish loading the remaining five back onto the truck.

The tailgate came down with a slam and I checked my watch, hoping the boys had made the school bus.

"What time is it?" Dennis called over the squealing of the hogs.

"Eight-twenty," I answered with a sigh. Time to start the day's work.

In the last two weeks of February and early March of 1989 we spent about seventy-two hours sorting, moving, and loading hogs. We sold one hundred and twenty-five head at an average price of thirty-five dollars per hundred weight, which is very low—about thirty-five cents a pound.

# Chapter Eight

On our farm, spring is a time for me to jump-start my muscles, which—after a relatively sedentary winter except for sorting and loading hogs—rebel at the demands of fieldwork. And to worry about my weight.

My ancestry is Irish, and I have inherited not only reddish brown hair—fast graying—but the Irish build—short and stocky. According to my doctor's chart, I have been forty to sixty pounds overweight for nearly twenty years. He assures me that my weight is normal for *me*, that there is nothing for me to worry about, but I do. So over and above the physical activity entailed by my job, I try to exercise.

Although I have prominent leg muscles that I developed as a child when I rode my bicycle endlessly, I am not a runner. Running hurts my lungs and jars my ankles. I hate it. But I love to walk and do so at every opportunity. My problem is finding the time. I try to get up earlier in the morning a couple of days a week to take a long walk—two miles at least—down the county road. I have tried walking in town in the municipal park while Kevin takes his swimming lessons at the community pool, but there were so many other people around and so much noise that I couldn't enjoy it. I am used to solitude and like to walk for peace, to talk to God, to find myself.

With each of my pregnancies I had gained about twenty pounds, which I've never been able to take off. After I gave birth to our son Kevin by Caesarean section I stayed home with the baby, breast-fed, and gained more weight. I was carrying an extra sixty pounds around when I became pregnant again four years later. I had a difficult pregnancy, was hospitalized for hemorrhaging, and confined to bed for several months. My baby was stillborn. And I found myself eighty pounds overweight. I could not run across the fifty-foot frontage of our house without gasping for breath.

Then I started helping Dennis on the farm. Since then my body has changed drastically. I used to have saddlebags on my hips but they are gone now. My legs are strong and firm. My upper arms used to sag. Now they are hard as rocks with some muscle definition. And my bust has increased a full size. But I still wish I could lose weight. I have tried all too often and failed. I have tried starvation and every diet ever described, plus a few I've invented, but each time I lose a few pounds I gain them back, and fast. I do not think I will ever bring my weight below one hundred and fifty pounds (which my doctor says would be my ideal weight even though I am so short.) I've concluded that it's not worth trying. My clothes fit and I have more physical energy now than I had when I was thirty.

Despite my weight problem, I pay careful attention to my appearance and I think my attitude is typical of the men and women in this area. I think most urban and suburban dwellers picture farmers as ruddy-faced individuals wearing dirty coveralls, their fingernails filthy—with wives to match. I don't know any farmers who fit that description. The only uniform thing about their appearance is the number of caps each owns, wears, and displays, each a gift from some agribusiness whose logo appears on the front just over the bill of the cap.

I once did encounter a farm woman dressed in a faded cotton dress worn over men's jeans with the legs rolled up, with men's high rubber knee boots, her hair scraggly, her face wrinkled and leathery from sun exposure, making her look much older than her age. She was strong but round and flabby and sported several jiggling chins.

I was so horrified by this apparition that I vowed I would never look like her. And most of the farm women I know are just as meticulous about their appearance although our general rule is the less makeup the better and we favor easily styled hair that we can pull back from eyes and face for safety's sake.

Although jeans are the foundation of my wardrobe I own several skirts, to wear to church and meetings, and a few dresses. But the only dress I like to wear is the one I sewed three years ago to wear to Jack's Confirmation and have worn to Kevin's first Communion and Jack's high school graduation. It is a full-skirted blouson made of polyester and cotton fabric in black patterned with large white lilies. It has a wide black belt and I wear it with black patent leather heels and a matching purse.

I love to sew and have even made Jack and Kevin pajamas and jackets on my machine. But now I have little time for sewing and I will need another occasion before I manage to set aside the time to make myself another fancy dress. Maybe this time it will be yellow, which is my favorite color and the accent color I've used in decorating our home. It also happens to be one of the few colors that Dennis can see well.

What I do collect is sweaters. I have so many I can't count them. But my favorite top is an oversized white cotton T-shirt with Old Glory emblazoned across the front in brilliant red, white and blue, beneath which is the message MADE IN THE U.S.A.

Farm women, and men, have to care for their skin and eyes not just out of vanity, but for their health's sake. When I started driving a tractor I was not accustomed to using moisturizers or sunblock. My skin burned, peeled, and burned again and its texture began to change. It became dry, and little wrinkles started to appear. Dennis's sister Alice asked her Mary Kay consultant to pay me a visit, and I began to use cleansers and moisturizers on a daily basis and sunblock every day we work in the fields. And in addition, when I work out of doors I slather on mosquito repellent and poison ivy medication so frequently that Dennis gets annoyed.

I have shoulder-length hair, which is longer than that of most of my women friends but I have a good hairdresser who gives me a trim every six weeks and a permanent when she can talk me into it. When I am working I am careful to pull it back or braid it. I also have to avoid wearing jewelry while working. I don't wear any rings—not even my wedding band—I keep my old watch in my pants pocket, and I do not wear earrings. The danger of having the post get hooked on something, resulting in a torn earlobe or worse, is too great.

Of course, my personal spring overhaul consumes only a fraction of the attention that must be given to work. By mid-April Dennis has usually planted some corn and made a list of repairs needed on the discs, planters, and tractors. Bearings for these machines must be greased and replaced on a regular basis, and since the planter has separate settings and attachments for corn and beans, it takes several days to make it ready for work. In spite of our maintenance activities, breakdowns seem to occur all too frequently, and it seems the worst breakdown will occur when it is least expected or needed. Virgil and I take turns making the drive to get the necessary new parts, in a hurry, so we can go on with our work.

I try to plant a good-size garden too, and I'm pleased if I can get in broccoli, cauliflower, spinach, potatoes, onions, and beets before the fields are sufficiently dry to allow us to move equipment into them. But it seems my primary task in planting season is to run errands and to move equipment. I spend a lot of time on the road.

Our corn planter has four seed boxes; each can be filled with one forty-pound bag of seed. The two fertilizer boxes each hold seven fifty-pound bags. The fertilizer is spread more rapidly than the seed drops from the seed boxes, so far more fertilizer refills are required than seed refills. Refills are often my job.

I also spend a fair amount of time in the spring on a tractor. In years past Virgil would go into a field first to break up the soil with the larger, more powerful disc and he might have had to disc over that field five or six times. Then I would follow with a smaller disc to smooth out the ground just ahead of Dennis driving the tractor pulling the planter. Since we have acquired a new, larger tractor and a big new disc we usually manage to do enough in the fall so that we can reduce the time we spend preparing the ground for planting in the spring.

We had very dry weather in the spring of '89. Early in April the temperature was already in the high eighties. Until we got a substantial shower—at least an inch of rain—there were places where we couldn't plant. I often had to run the shallow disc a scant eight rows (two rounds) ahead of Dennis on the planter so he could get the seed in before the upturned soil dried out. (I was sun-blistered three times that spring—in early April just to my elbows because I had been wearing a sweatshirt; then on my left arm, the side exposed when I spent three hours discing one morning; the third time was on the last day of planting when I was on the tractor for four hours in the late afternoon and early evening. Of course, each time I was driving Goliath, our only large tractor that has no cab. Since then I have been very careful to apply sunblock before going out to work even on cool or cloudy days.) Despite all obstacles we finished planting in record time on June third.

Summers here are short or seem so, as everything has to be done at top speed. The fact that we were having a drought didn't seem to bother the weeds in my garden. But Jack was taking gardening as a 4H project and so he did most of the weeding and hoeing. Dennis and Kevin did the potatoes and beets. But we all had a hand in botching the green beans.

First, Dennis planted three rows of bush beans early in the spring, but the rabbits ate them all before the plants were six inches high. Jack put in a fourth row but he must have planted too deep because only three or four plants emerged in the twenty-foot row. So Kevin and I decided to show them both how to do it right. For two evenings we worked hard hoeing the ground, preparing to plant, then carefully setting the seeds in the ground several inches apart in two perfectly straight rows.

In summer, we use more water in the animal buildings: We do more cleaning and the pigs need more water too. We have a complex irrigation system that channels the water from the primary lagoon to an overflow lagoon, but it involves above-ground pipes and is so complicated that we don't use it unless absolutely necessary. So we were relying on an earth dam to channel the water from the lagoon into the ten-acre field at the bottom of the hill.

The night we finished planting our beans the dam broke. Our beans were drowned.

But the remainder of the garden sure was fertilized! We grew tomatoes the size of cantaloupes, cantaloupes the size of watermelons, and zucchini two feet long. I canned over one hundred quarts of tomatoes plus twenty-four jars of chili sauce and I swore I would never plant another tomato. I canned and pickled and froze garden produce until I was chopping and slicing in my sleep. But the only green beans we got were those I picked in Dennis's parents' garden while they were on vacation.

# Chapter Nine

There are many county and state agricultural fairs in the summer at which large farming and agri-related businesses show and sell purebred livestock. We are proud of the quality of our hogs, but we don't raise animals for show. We choose our breeds carefully, but rarely raise a purebred hog. So the only fair we attend is the Gasconade County Fair held in Owensville the last weekend in July.

By the end of July all the crops are in the ground, sprayed

and growing. Dennis tries to schedule breeding so the sows don't farrow during July and August, the hottest months of the year. So there is nothing to prevent us from attending the fair every day and enjoying ourselves. It's a relatively small event, mainly attended by local people, perhaps two or three thousand in the four days that the fair is held. Its main attractions are an antique tractor pull held on Wednesday evening, opening night, and the 4H-FFA livestock auction held on Saturday, the final day. Jack has entered butcher hogs he's raised himself in this event for five years running. In between there are pig scrambles, sack races, and pedal-tractor pulls as well as a parade on Thursday evening that Kevin loves. Celebrities entertain; we have had 'Exile, Dave and Sugar, Sha-Boom, Jimmy Dickens, and Cissy Lynn. But we're not in the same league as the big fairs nearby that draw crowds of three to five thousand a day, like the Montgomery County Fair, which is due north of us, or the Washington Town and County Fair, due east, where there are simultaneous truck and tractor pulls and appearances by big-name entertainers like Loretta Lynn, Conway Twitty, Charlie Daniels, Lee Greenwood, and Alabama. And these fairs are just stepping-stones to the big state fair held in late August in Sedalia, Missouri.

Once I was a contestant for the title of Warren County Fair Queen, but that was a long, long time ago. The Gasconade County Fair does not have such a contest, so Miss Missouri State Fair isn't going to come from our area. Washington, Boone, and Montgomery counties will send their queens on to Sedalia; we'll just send our produce and our pigs.

The exhibition halls are permanent—white-sided two-story structures in which, each year, 4H exhibits are displayed. Competitions are judged in sewing, woodworking, cake decorating, baking in several categories, home-cured hams, garden produce, and dozens of crafts, changing as interests do. Many people seem to try to submit items that will travel on to the state fair competition, but it seems to me that some of the entries in the home economics building—quilts, crochetwork, knitting—appear to be the same every year. Local businesses also display home-repair and improvements items, farm implements, insurance information, religious goods, and lots of T-shirts, jewelry, and balloons.

The ground story of the 4H building holds the garden and farm produce exhibits: tomatoes, potatoes, squash, eggplants, cantaloupes, watermelons, pumpkins, field corn on the stalk, soybeans still on the plant, and full bales of sweet, fresh-smelling hay. There was a time when the competition in this building was the focus of the fair but interest seems to have dwindled. These days the two rows of tables down the length of the fifty-foot building are barely half-filled with produce.

After the sun goes down, the food, beer, and cold drink stands are crowded. Here the picnic tables are set up and customers can buy fish sandwiches, funnel cakes, hamburgers, and ice cream cones.

Beyond the food area and the displays is the midway lined with game booths where customers try to win pictures or stuffed animals, the cotton candy stands, and the curly fried potato and lemonade purveyors. And then there are the rides:

the Scrambler, Tilt-a-Whirl, Paratrooper, and Octopus, the fun house, carousel, and Ferris wheel.

The Gasconade County Fair is like a big block party. There are few strangers in the crowd. Anyone you don't know will soon be introduced to you. The adults sit and visit while the children run off to play and visit on their own. But I am much more likely to lose Dennis at our fair than one of the kids. There are some hazards, of course. It is generally hot and dusty, and there are some mighty small kids trying to show two-hundred-fifty-pound hogs and two-thousand-pound steers in the show area.

The livestock auction draws several hundred people. Some bring their own chairs, others occupy the portable bleachers that are set up for this event under the permanent fiberglass roof of the open-air arena, which is located close to the cattle, hog, and sheep barns, each a long wooden building filled with portable pens.

The auction itself tends to be a lengthy affair as more than one hundred young entrants show their market lambs, barrows, and steers. Four hours is about standard for the entire event.

One year Jack was showing so we were anxious to find ourselves seats for the auction. The heat was stifling in the portable bleachers. Hardly a breeze reached us. Some one hundred and fifty members of the audience were sitting in the bleachers or on lawn chairs that stood beneath the fiberglass roof or were scattered around the dirt-and-sawdust-covered area in front of the iron bar fence around the show ring. The auctioneers, high on their own stand, were opposite from us.

We sat there nervously as first the steers were sold and then the hogs. Jack, who is a free spirit, always wandering the fairgrounds even while the auction is taking place, was nowhere in sight. Finally, I spotted him, standing opposite us with a group of other kids. I needn't have worried. This 4H-FFA auction is a money-making venture for the participants. It provides them with savings either for college or toward the purchase of their own cars. Some will earn triple the market price for their hogs or steers. Not one of them would risk missing out on this chance, the culmination of six months of work.

Each participant writes letters to local businessmen asking them to come to bid and ultimately buy. The recipients respond in hopes of earning goodwill. Local newspapers write them up and customers are attracted. Jack had spent hours painstakingly writing out a half dozen such letters, a task that was for him the hardest part of his six-month project.

I glanced around and saw many familiar faces. A couple of the volunteer auctioneers were family friends, and I knew many of the other parents. I also saw some of the agribusinessmen to whom Jack had written. I hoped their presence showed the respect they felt for Dennis and our farm operation. But no doubt others from our area had written to them too.

Jack was now near the entry gate with his barrow. His hog was the next to be sold.

The barrow was in no hurry to waddle to the center of the stage. He nuzzled the sawdust and dirt beneath his feet at his leisure. Jack walked behind him with one hand held out inches above the young hog's back ready to nudge him should he try to lie down or threaten to stray out of the bidders' sight.

The auctioneer introduced them over the intercom: "Jack Kottwitz from Hermann, Missouri, showing a Large White two-hundred-and-sixty-pound barrow. He's a good-looking hog, folks. One of our blue-ribbon winners. Let's start the bidding at fifty cents."

The arena came to life.

The auctioneer rattled off numbers so fast the untrained ear could barely follow him. Just outside the show ring four of the auctioneer's helpers strode around, each in his own territory, encouraging bidders and shouting out a number each time a potential buyer agreed to bid.

The auctioneer called out bid after bid. They came from the left, the right, the left again, then the center.

My head was spinning. Yet it seemed I scarcely had time to blink before the auctioneer smacked his gavel down crying, "Sold! Two ninety per pound!"

The hog had been bought by someone far to our right but I had missed hearing his name. And Kevin hadn't heard either. But Jack and Dennis knew who the buyer was and they exchanged satisfied smiles as Jack, his hand pressed to the barrow's spine, ushered him from the ring.

Ten minutes and several barrow sales later, Jack climbed up the bleachers from the rear to tell me that the John Deere dealer had bought his barrow.

It remained so dry that year that we began to combine in mid-August. The crops were far better than we had anticipated, but not as good as they could have been had we been blessed with normal rainfall. We had been lucky that the one substantial rain we got occurred on the July Fourth weekend right when the corn was beginning to tassel; several inches of rain at this point proved our salvation. But early in September Dennis had to stop combining corn and switch to soybeans as they were so dry they were popping out of the pods while still in the field on the plant. As a result, by the time he switched back to combining corn we were already trying to sow wheat. It should have been manageable, but rain arrived just when we didn't need it. So we didn't get as much wheat sown as we had wanted, and what we did sow didn't grow as it should have. This was the moment when Dennis decided that, since I'd mastered tractors large and small, it was time for me to learn to drive the combine.

Dennis recalls that I didn't want to learn to drive it. He's wrong. I considered that learning to drive the combine would be one of the greatest accomplishments of my life. I had spent three years greasing underneath its corn-head or washing its windows while I was standing on top of either the corn-head or bean-reeler five feet in the air, or running alongside it to clear away weeds before they clogged it up: I thought it would be great to work inside our combine, for a change.

This machine was new. Our old combine would have been ruined by having the dirt and grit of the 1986 flood residue run through it if it had not already been on its way out long before that. It seemed to me that as far as trying to repair the old combine, we had already reached and passed the point of diminishing returns so I had come down in favor of buying a new one.

This combine is a relatively small machine as combines go; its attachments only clear out four rows at one time. But it still stands taller than two pickup trucks stacked one on top of the other. The front wheels are five feet high—and so am I. I can curl up inside the rim of a mounted tire and take a nap. So even with power steering and power brakes it's not easy to get it to take corners or to stop on a dime. I have nicknamed it Dino because when I stand at the end of a field and watch it approach it reminds me of a dinosaur, a big brontosaurus, lumbering between the rows.

Our combine's formal title is 1985 Deutz-Allis F3 Gleaner. The walls of its cab are almost entirely window glass, floor to ceiling. They have to be because the driver must be able to see in three directions, but these walls are a job to wash, a chore that must be done almost daily because of the dust and clutter kicked up from the ground by the headers. An electronic dash located overhead lights up continually with all sorts of messages concerning the operation of the machine. The driver is supposed to monitor this as well as steering and watching the header, directly in front of and beneath the driver's seat, to observe everything that is feeding into the machine. The cab is air-conditioned fortunately; otherwise, the driver would suffocate from heat or choke on the dust and debris flying out of the headers.

The motor is located just above and behind the grain bin, directly behind the driver. There's a little rear window through which the driver looks to check how full the grain bin is. The gearshift is located on the far wall to the right instead of on a column; to shift gears you turn a knob clockwise or counterclockwise. It's not much like driving a car. The header is controlled by a lever; to work it all you have to do is push a button. But with all the driver has to do to control all the functions that have to be performed and the sheer power that could be unleashed, the operation of the combine had always terrified me.

But once he decided that it was time for me to learn, Dennis didn't hesitate. He got into the cab with me, grudgingly adjusted the seat so I could touch the brake pedals with my toes, lowered the steering wheel so I wouldn't hit my chin with it, and said, "Okay, put it in gear."

I finally found the gearshift knob and steered Dino into the field, pushed the little button, and—dropped the tips of the corn-head clear into the ground.

Dennis yelled, "Lift it up! Lift it *up!* You'll tear up the chains!"

And he leaned over and hit the button without even looking, at once lifting the header to the exact height he wanted. I'd had to let go of the steering wheel when he leaned across so we veered three rows off our course. For this, I caught hell again! I made it back on course and drove to the end of a long field without major mishap. Some of the ears of corn even ended up in the header. But then we had to turn around.

To turn Dino in this field I had to back up and turn, back up and turn, back up and turn, back up again until the points of the header faced exactly between the rows in the return direction and travel back to complete the round. Dennis did far more steering than I on this first attempt. But at the end of this round Dennis called out that the bin was full.

"Okay, Mim. Now, just unload it."

Right! Dino has an automatic auger, so I wouldn't have to leave the cab to unload. But first I had to drive the combine alongside the farm truck, which was already three-quarters full, and position the auger so that the corn would land right in the front center of the truck bed. I didn't quite make it, so Dennis had to park for me.

The lever that releases the auger is directly behind and beneath the seat. Dennis can reach down and pull it up with his left hand. I had to get up, back out the door, and use both hands to force the lever up; to give me room, Dennis had to vacate the cab by backing down the stepladder over the front wheel.

But I succeeded in performing this operation, so he hopped down and announced cheerfully that he would drive the filled truck to town while I loaded the other one.

"You're going to leave me alone to run this thing?" I asked.

"Why not?"

"And what if I hit a tree?"

"The trees are all the way over on the edge of the field."

That was not a good enough answer for me. I climbed down. I was not ready to take the blame for all the things I feared would go wrong if Dennis left me alone with Dino. I wanted to drive Dino, I was sure I could—but not sure enough. Not yet.

# Chapter Ten

No farmer can survive without friends.

I suppose we all have acquaintances, people with whom we share different aspects of our lives. I know I do. And there are relatives, some of whom are friends too. I have an aunt who has helped me and supported me with her love and wisdom all my life. I did not know Dennis's sisters before I met him, but now they are my sisters too, as his parents are my parents. But there are not very many people who have made a significant mark on my life other than kin.

In a rural area like this one, there are fewer people in a forty-mile radius than there are in a single block in New York City. So most everyone knows one another, and this goes generations back: who to trust, who can be depended on, who to avoid. Few people are actually "blacklisted." But an argument or misdeed is remembered long after the people involved have died. Fortunately, most people are also very kindhearted and more than willing to forgive.

I am a newcomer to this community, and it has taken me quite a while to meet people Dennis has known all his life. I still get some of them mixed up but already many are very special to me. The man who owns the packing house down the road does a lot of work swapping with Dennis, and his wife is my good friend. I remember when we lived on Grandpa Hugo's farm how he dropped everything to locate Dennis for me when our outdoor well house caught fire.

We see Pete, our feed dealer, every week because we do a lot of business with his company. He is a friend. He helped us move into our house eight years ago. Time and again he has helped Dennis to castrate or move hogs, or to make some repair. If we need an item, Pete keeps his eyes and ears open for us, and usually manages to locate it. We socialize with Pete and his wife; we enjoy their company.

Paula and her husband, Ted, are not farmers but Ted comes from a very large family and grew up on a farm. Farming is still in his blood. Throughout the spring, summer, and fall, he comes by to lend a hand where he is needed. Together, he and Dennis may plow or disc until the wee hours after they have each worked a hard day. Paula and Ted have not only helped with work, but have given us water from their well, electricity, and tools, and have loaned us a building in which to make repairs and offered us the use of their storage shed for trucks and wagons to shelter them from the rain.

A couple of years ago, one of our landlords baled straw from the wheat stalks the combine left behind. Dennis and Jack got in two loads but had to leave the rest behind to do other work. A few days later, Paula and Ted and their daughters surprised us by bringing in the third and last load and delivering it to us in their pickup. Another time we stored a truckload of wheat kept for seed in their storage shed. When we came for it, a storm swept down on us and we couldn't move the truck for fear of wetting the seed-wheat. So we had to wait. Paula came down to chat. Then Ted arrived from work. Out came the beer and picnic bench from the porch to sit on. I don't like beer, so Ted went to the house for a bottle of wine for the women. Then Ted and Dennis decided to check out the deer, leaving us with the wine bottle. Time flew by. When the men returned at long last, neither Paula nor I was in any condition to drive anything anywhere.

That's not our usual style at all. Paula is a blue-eyed blonde about my height who has a slightly cynical but still fantastic sense of humor. I first met her through her sister-in-law, my dear friend, Bonnie.

When Jack was only seven I volunteered Sunday mornings to help teach the class in Confraternity of Christian Doctrine to the first and second grades. I hadn't lived in Hermann long and I knew very few people there. But in the short time that I'd been married I had already given birth prematurely to a son who died, a baby for whom Dennis and I still grieved. When Bonnie volunteered to be my assistant teacher, she also became my first friend in town, one whom I needed badly. She was a wonderful helper. Indeed, she was far more than an assistant to me but it took her three more years before she summoned up the courage to teach a class on her own. I became pregnant again, and although there were complications and I was hospitalized for weeks, our son Kevin lived. Four years later, I had a stillborn child and my doctor told me that although I would continue to conceive I could never again carry a child long enough for it to survive.

I was devastated. My main sorrow was an inability to have more children but I was also faced with a terrible dilemma. How could I endure a future of repeated, doomed pregnancies? But how could I stop loving my husband? My parish priest had no answer for me.

I believe in my church as a holy institution but I also realize that, as a church controlled by man, it has serious faults.

I believe that life begins at conception and continues after death. And I believe God is a kind, loving master who does not interfere with our personal quests. Like a loving parent He must stand aside to watch us grow and become successful on our own. He watches, often sadly, as we mortal human beings exercise the free will He bestowed upon us. He cries when we cry. He laughs when we laugh. And He is always there, waiting on the sidelines, for us to acknowledge Him.

Therefore, I am also convinced that the officials of the Catholic church often lose sight of their own immortality in their efforts to deprive a woman of the right to exercise her God-given free will with their rulings on divorce, birth control, and abortion.

Certainly it is wrong to remain with a spouse who beats, deserts, or degrades you and your children either physically or psychologically. If two people do not love and respect one another, how can they pass those traits along to their children? And is it not a greater sin to chance pregnancy knowing you have no means to support the new life and it would endanger the comfortable home and hopeful future of the children you already have?

I do not condone abortion. I would never seek an abortion nor would I advise anyone else to seek one. Neither do I have sympathy for the women who, for reasons of money, self-esteem, or lifestyle, seek abortions. I do not condone the violent protest methods of either right-to-life or pro-abortionists but neither do I want to see a return to the back-alley butchering women suffered in the past.

I have been in a life-threatening situation during childbirth and I know, had the decision been necessary, my husband would have chosen my life over the baby's. I'm glad we were not faced with that decision. I also remember the feeling of despair that engulfed me when a priest offered to administer "the sacrament of the sick" to me as I lay in labor with our son who died. When I told him it was not my life that was in danger but that of my baby, he turned away sadly, his own despair written all over his face. Ironically, my unborn, unbaptised infant was not worthy of the special sacrament to pray for its life.

In the end, I solved my dilemma by having a tubal ligation. And I took whatever guilt I was supposed to have felt (I call it confusion) into the confessional only to have the priest there tell me I was not wrong to do whatever I had to do in order to survive, for the sake of the children I already had.

This confirmed my belief that each person must make his or her own choice, do whatever he or she must, in order to survive. So long as the decision does not harm others; so long as he or she can remain at peace with God, his or her fellow man or woman, and himself or herself, it is no one else's business . . . not any church, certainly not any government.

Our government was founded on the belief of separation of church and state. So if the government plans to continue to uphold this belief, it should not adopt laws as a convenience to itself or any religious, ethnic, or political group. Like the officials of many churches, too often we see our government officials likening themselves to God and infringing upon man's God-given gift of free will.

Throughout it all—months and months of mental anguish—I received more understanding, guidance, and counseling progress from my women friends who found themselves facing the same dilemmas (to a greater or lesser degree) than I ever have from any clergyman or doctor.

Bonnie was there for me when I needed her all during these hard times. We taught CCD and planned our lessons together, shopped together and talked and laughed—and cried—together. It was Bonnie and Paula, together with JoAnne—another friend I met while teaching CCD—and Susan, my girlhood friend (now, a registered nurse specializing in obstetrics and gynecology) who gave me comfort and support.

Bonnie, who already had three children, was concerned about birth control although she was also a devout Catholic who had converted when she married. And JoAnne, who had undergone a hysterectomy after the birth of her second daughter, when she was in her midtwenties, understood my feelings of anger, resentment, and failure. For months I truly hated pregnant women and young women with healthy infants. I thought I was uniquely depraved until JoAnne confessed that it had taken her years to overcome similar feelings.

The first year we taught together, Bonnie and I attended the annual Religious Teachers' Institute in Jefferson City, and we kept this up every year for a time. The institute is held at the end of October and was often the first chance we had to see one another since the preceding May when classes ended. In the quiet intimacy of the car as we made the one-hour drive, we would catch up.

The meetings would open with a welcoming Mass. Then we would split up to attend workshops, meet again for lunch, and talk over what we'd learned and ways of implementing new ideas. During one final session held in St. Joseph's Cathedral, a round edifice that resembles a vast auditorium more than a traditional church, several hundred assembled teachers were instructed to reach out and hug those on either side of us. Both Bonnie and I were embarrassed. Hugging strangers had not been part of our upbringing. But talking it over on the way home, we decided the gesture was really quite meaningful. We had made friends at the annual institute, although we might not see one another from one year to the next. And friends should love one another. Indeed, we wished our parishes were more open and accepting of others, though they might be strangers.

Paula began teaching CCD and she too came to Jefferson City with us. JoAnne taught too, but never made the annual trip, although some of the other teachers have done so. Still, we all kept up our separate friendships.


The year ended on an upswing. Hog prices were good enough. Despite the dry weather, crops were very good. So good, in fact, that there was money left over after all the bills were paid to pay for ceiling tile and a new tub for the bathroom. Then at Christmastime a cold wave hit, bringing bitter weather for several days before the end of the year. Dennis had gotten a deer during the season, as did our brother-in-law and Virgil, so deer steaks and sausage were plentiful.

We spent Christmas Day with Dennis's parents in their home. On New Year's Eve a small group gathered at Ted and Paula's—Dennis and I, and Bonnie and her husband, Jeff, and Ted's brothers and their wives. Paula was pregnant so she wouldn't drink, but I sampled a little of the homemade wine. Bonnie was very quiet. We thought she was having another one of her headaches, headaches that we—Paula and I—jokingly blamed on Jeff.

And then at midnight, Ted went outdoors to shoot his rifle into the air, to bring the New Year in. And it came.

On January fifth, Paula was laid off from her job due to company cutbacks. That same day, the cause of Bonnie's recurrent headaches was diagnosed as a large tumor located near the optic nerve.

Two days later, before a CCD class, Paula told me that they were going to operate on Bonnie the next morning. The boys and I had just gotten home when Dennis's mother called to tell me that Dennis's only living uncle had just had a second heart attack and the prognosis was not good.

Dennis was chasing down one of our landlords in order to pay him our annual rent and he didn't come home until early that evening. Though I was trying to hide my tears in the steam rising from the sink where I was washing dishes, he knew as soon as he saw me that I was upset. I had to tell him the bad news.

Dennis's uncle had bypass surgery (and has recovered completely), but Bonnie was found to have a cancerous tumor, not all of which could be removed. After the operation her sight and speech were affected and she had to decide whether to undergo treatment that might prolong her life for another three years or so, or just to make the best of whatever time she had left. Bonnie was not physically beautiful; she was short and much too heavy for her height. But she possessed an inner beauty; her warmth and spontaneity made everyone her friend. There was no vanity, no arrogance in her. Indeed, she was so shy that she could never bring herself to return a mistaken purchase to the store. She just put it aside rather than bring it back for a refund. But her bravery now was extraordinary. She was determined to fight. She resolved to undergo chemotherapy.

Bonnie was only thirty-seven and her three children were fourteen, twelve, and eight years old. Now that she knew she would not live long enough to see her daughters marry and have grandchildren, her dearest wish was to see her youngest son make his first Communion. Immediately after she was released from the hospital she had her lawyer draw up a living will so that her family would not have to suffer any longer than necessary by having to watch her hooked up to a machine long after hope of life was gone; she had faced that possibility too.

Paula, several months pregnant, spent the days with Bonnie, who in addition to loss of vision, was now subject to grand mal seizures. I went to visit when Paula had to leave Bonnie for her own doctor's appointment. Dennis thought I should wait until Bonnie was better, but I didn't want to put it off. I was glad to see her happy despite the fact that I had to speak before she could recognize me and her speech seemed slurred and halting.

"I'm going to beat this," she repeated several times. "I've got a lot of living to do yet."

As I hugged her goodbye I told her, "The next time I see you I want you to be completely well again."

It seemed to be a season of misery. Wintry weather and all the ailments flesh is heir to.

I got home to find that Dennis's blood pressure was climbing again. It reached 200/112 and he had to go on medication. Dennis becomes extremely ill when his blood pressure gets high; I had to take him to the hospital in Hermann to the emergency center before the medicine started to take effect.

Dennis is never very patient but he seems to undergo a personality change when he isn't feeling well: He takes offense at everything and seems to be beyond pleasing. His appetite deserts him too. Over the course of the next year and a half, while his medications were being changed without much effect on his blood pressure, he lost forty pounds. He had been overweight before, but this didn't help his disposition.

In February, Jack broke the thumb on his right hand. For the next six weeks he couldn't even button his jeans, much less do his daily feeding chores. So we switched jobs. He managed most of the housework pretty well except for the dishes. But the kitchen floor was never cleaner because every night when I came in from feeding, I had to mop up the soap suds and dishwater from it.

We began to repair the auger in the finishing building that had broken down completely the previous year. It extends the length of the building a scant foot beneath the ceiling and is held in place by bracing attached to the roof beams. Dennis seemed unable to comprehend my inability to reach the rafters without a ladder, as he could. My assistance was limited to standing with a bucket on my head to catch feed as he tested the auger and to holding tools and handing them up to him and chasing pigs out from under his feet.

While the auger was being fixed, we had to keep putting in feed through the windows, raising and lowering the pulley by turning a handle. I began to feel twinges of pain as I performed this task, but when I stopped the pain stopped, so I ignored it. Then one night as I lay down, pain began to travel from my breast around my rib cage to my back, shoulder, and neck. The next day, when I tried to use the pulley it slipped from my hand and I was almost knocked in the head by the rapidly unwinding handle. Then, a few days later, I couldn't catch my breath as I was folding laundry. Finally, I too had to see a doctor.

I had stretched one of the lateral support muscles circling under my arm until it had lost resilience so my diaphragm could no longer expand properly. A little more neglect and I would tear the muscle and need surgery, the doctor said, but if I stopped using that arm altogether, I would develop a frozen shoulder and lose its use permanently. Well, I've had enough operations; I decided to tough it out—while being careful. I was sent home with ibuprofen, a single exercise to perform once a day, and the warning that, in the six weeks that the muscle would take to heal, the pain would get worse before I got better.

The next day when Dennis asked for my help filling the feeders in the hog building, he wouldn't believe that I was so "weak" as to be unable to do this simple job. But as soon as I turned the lever on the auger, my right shoulder felt as if I had torn it in two. I had to use my other arm to lift my dangling right arm. Dennis, who hadn't thought there was much wrong with me except for a few aches, saw I was really hurt and from then on raised and lowered the auger himself.

By the time he asked me to help him unload feed in the outdoor pens, the agony had worn off so, without thinking, I reached out with my right hand to open the gate—and doubled over in searing pain, my right arm curled at the elbow at my waist. At the house, my sons had to help me take off my coat. And for the next three weeks I was one-handed, and mad as hell, as I spent the nights moaning or walking the floor, keeping Dennis awake to his annoyance, or dragged myself around drugged with ibuprofen so I could stand the pain, unable to carry buckets, open gates, or shut doors with my right arm so I was almost useless outdoors, and nearly as useless in the house.

Jack was still in a cast so, as I retained the use of both hands, I had to keep on doing the work in the farrow house for him. We could haul feed through that building in a large three-wheeled cart and I am left-handed, so I could scoop feed from the cart with a plastic scoop easily enough. But there were other chores that were harder for me.

It seemed as if every sow we owned was having piglets and delivering litters so large that some of them could not produce enough milk to keep all their babies alive.

We have a small auxiliary farrowing building for overflow.

We can house ten sows and two hundred piglets in this twenty-four-by-fifty-foot building, which has two wings in which there is steel-wire and cross-bar flooring over open sewage pits separated by two central rooms with concrete floors where feed and other items can be stored. But a few years earlier when Dennis began to cut back on the size of the breeding herd, the building had flooded when the sewage system clogged. And although Dennis had cleaned out the drainpipes and they were working, he'd never cleaned up the mess. In addition, the previous winter the water pipes had frozen and burst and they now had to be replaced.

We were faced with a monumental cleanup task, complicated by the need to clear out accumulated junk that was piled to the rafters in the central rooms and to repipe the rooms and attach long hoses down to the drinkers in the pens so the sows and piglets would have a continual supply of drinking water. And I had one usable arm.

Simultaneously, the big farrow house was filled with piglets needing their teeth and tails cut, their shots, heat lamps, and also daily mucking out. Dennis can cut a piglet's tail and teeth, and give it its shots in less than two minutes. I had mastered a technique of my own, based on his movements, though at a much slower pace—but two arms were definitely needed. So Kevin and I worked out a method using both of his hands and arms, my left arm, left hand, and some of my right hand. He would catch a pig and I would take hold of it with my right hand, grasping it just behind the head. Usually, the pig would squeal so I could get the clipper into its mouth to nip off its tusks. Then Kevin would turn him around so I could seize his hind legs. I would inject the piglet in the hip rather than the shoulder as Dennis does. So with Kevin taking the clippers and handing me the syringe, we managed.

After nine weeks I regained the normal use of my right arm, but I still had trouble winding the pulley on the feed grinder auger.

Trees had to be cut again. Some had died. We had to make room for new growth or have a hillside of dead, topless trunks, to greet the spring.

The day before Palm Sunday, Kevin helped me defrost and clean out our twenty-year-old eighteen-cubic-foot deep freeze. We spent the morning emptying it of packages of meat, scrubbing it out, and restacking the contents in an organized way so we would be able to find what we wanted at a glance. Four days later, I discovered the deep freeze was not working; the meat was thawing. By the end of the week, bloody water and sticky ice cream were mixing and melting all over the floor. I cleaned it up once and then, at Dennis's suggestion, we chewed enough gum to pack the drain hole and keep the freezer from leaking all over the floor. But we lost the entire contents of the freezer.

Then my washing machine went kaput so I had to haul the laundry to town to wash it.

We celebrated Easter with Dennis's parents. On April 20, Paula had a healthy baby boy. A few days later Dennis and

Virgil began to plant the first of the corn. For two days, all went well.

Then a month of rain began.

And Bonnie died.


May first is my birthday; Bonnie's funeral was the next day. Dennis refused to go to the funeral home with me, with an excuse—typical male denial. Fortunately, my Aunt Liz was able to visit me on my birthday and she helped to cheer me up.

"Why do people die so young?" I asked her.

"Why do people live to be so old?" she countered. She was about to celebrate her eighty-first birthday.

I went to the funeral with JoAnne. Her husband had ducked out too.

I miss my friend Bonnie. I guess I always will.

# Chapter Eleven

Like most women, I have become involved in the life of our community through my children. I started volunteering to teach CCD to first graders when Jack was only seven. Later I attended workshops throughout the state of Missouri for about eight years and today I am a teacher of religion certified by the Jefferson City, Missouri, diocese. I also participate in our community's life through 4H volunteer activities.

For a few years I have worked with a very supportive group

to put on a ten-minute skit in competition with other area clubs on 4H "Share the Fun Nite." In the spring of 1989 we put on a skit patterned after *Wheel of Fortune* that we called "Wheel of 4H." We won third place but still the kids were disappointed that after all our hard work we hadn't done better. The next year, I vowed, we would.

It seemed to us that in recent years the judges had been losing sight of the rules that prescribed a ten-minute length for the skit and the provision for all cast members of public speaking experience. Past winners had skits that ran too long and relied on a single narrator to do all the talking. And while no 4H theme was required, past winners had been clubs that had chosen 4H themes. We decided to tell of the difficulties in deciding on a theme and the problems in putting on a skit in an original ten-minute production entitled "Committee Meeting." That way, even if we didn't win, we would have delivered our message.

Our club was working hard. Our worst problem was scheduling—finding a time when we could all get together—followed by the persistent problem of actors who wanted to change their parts or rewrite their lines. At this time of year there were so many school events, and so much fieldwork, that some rehearsals were very sparsely attended. Nevertheless, the kids seemed to retain their enthusiasm even after I had started to feel qualms.

"Share the Fun" takes place in the community theater on Main Street in Owensville and is open to the public. The theater, a converted movie house, holds nearly three hundred people. But there is no "back stage." Instead, there is a small prop room and alleys out back where clubs are forced to assemble, dress, and apply makeup all at once and all together. On the night of the performance it is a noisy and confusing place.

We were scheduled to go on in the eleventh spot so I was able to sit in the audience for the first few skits. Then I had to gather our club members to go out behind the theater. As I headed up the aisle, one of our stars, a pretty girl of fourteen, stopped me cold with the announcement, "I don't want to go on stage."

I don't think she had stage fright; I think she was one of those never satisfied with her part and I was pretty annoyed.

"You made a commitment," I told her. "Now you have to keep it."

She never even came back to the alley, nor did another of our actresses who had a major speaking role.

So I gathered the three boys who were the remnant of the cast of "Committee Meeting" and broke the news to them that they were going to have to manage without the prima donnas. They quickly shifted lines around and retreated to the end of the crowded alley to practice, continuing to rehearse even while we were setting up our props onstage. They were perfect. Not one cue was missed in the course of the eight-minute performance; it seemed just as if the script was meant to be the way they performed it.

The culmination was Jack's appearance giving a demonstration of how to bake a black marble cake. When he gives a demonstration, a 4H member is supposed to be totally prepared. Jack had a complete cooking setup: bowls, pots, pans, utensils, and all necessary ingredients lined up before him on a folding table, but he gave a spoof cooking lesson, producing a cake made from baby oil, raw egg, plastic flowers, and dirty water as well as a cup of black marbles.

It was worth the trouble it gave me to clean up the stage after him, because our club was the first-prize winner and was awarded the trophy. And, of course, I was particularly proud of Jack.

Jack had been diagnosed with Type I diabetes mellitus, which makes him insulin dependent, when he was twelve. Sitting in the waiting room of the doctor's office he was so scared when he heard this verdict that he broke out with hives. But our doctor was wonderful.

He told Jack, "I was eleven when I got diabetes and I'm forty-seven now." And Doc J. is the picture of health.

"Diabetes isn't going to kill you unless you let it," he told Jack. "You'll probably never get over it because they haven't found a cure yet, but it need never be more than an aggravation."

It hasn't been easy, however. The disease is not something you can ever ignore. It is right there, demanding attention. The diet, which has to be followed strictly, is complicated, and the fear of insulin reactions, which occur seemingly spontaneously and require immediate attention and an adjustment of the diabetic's regimen, is constant.

From the outset, as soon as the diagnosis was made, Doc J. insisted that Jack attend a camp for diabetic children located near Columbia, Missouri. It is coeducational and has good facilities and a superb medical staff. After I visited, I had no qualms about entrusting Jack to the care of the counselors, although I would not be able to contact him unless there was an emergency. Still, I knew he was uneasy about being left there. Driving away that first time was one of the hardest things I have ever done.

The next summer, he hardly noticed I was going, he was so busy renewing acquaintanceships. Several of his fellow campers became real friends. And in his third year of camp, he started talking about going there, looking forward to it, in January. At the end of his stay, he was chosen Best Camper and invited to remain for the second session as a counselor in training. He had to refuse as he had hogs he was raising to sell at the Gasconade County Fair.

Jack had to drop out of Future Farmers of America after ninth grade because of a conflict between the class times for agriculture-related courses and college preparation courses in science and math. Jack has had a lifelong interest in biology, but he has been an honor student ever since elementary school and his ambition lies in the field of medicine. He raises hogs to sell at the fair to make money for his college fund.

Jack had been alone so often and borne so much responsibility on the farm that I had begun to think he could take care of himself and I had stopped supervising him as closely as I should have. And he had just gotten his driving license; he didn't want to be treated like a child with his mother hovering over him now that he was officially a young man. He came down with a virus and began to suffer severe insulin reactions, several times fainting unexpectedly. But when, after two days in bed, he seemed to be fine again I put it out of my mind.

When Jack was first diagnosed as a diabetic in the middle of seventh grade, his school had been informed of his situation, and by the time he entered high school I had had several conferences with the school nurse and middle school secretary about Jack's illness. I assumed the high school authorities also understood the problem. I assumed too much.

Dennis and I had been out, working with the hogs, and returned for lunch to find a message from the principal of the high school on the answering machine.

"We've had some problems with Jack in school today," he said. "Please call back at your earliest convenience."

Even before I dialed, I had a premonition. What the principal told me was that Jack had been belligerent in school and had called a teacher a bitch. He was consequently suspended for three days.

I was sure that Jack had suffered a severe insulin shock. He is a boy whose conduct had always been impeccable, to whom maintaining good grades mattered. This sort of behavior was totally unlike Jack. And I knew, all too well, that when a diabetic has an insulin reaction he may not only sweat heavily and feel disoriented and dizzy, but he can become violent.

Jack has thrown furniture about when he was having a bad reaction; sometimes he has been unable to communicate. When his blood sugar does not rise to a normal level quickly enough, he has something like a panic episode.

I also know, to my sorrow, that the most vivid memory Jack seems to have of his father is of violence resulting from his severe insulin reactions. John had suffered heart and kidney damage, and become blind before he died, and he was angry and frustrated for much of the time. He had been stubborn, impatient, and proud, and the knowledge that he was doomed to die so young had made him extremely difficult to live with in his last years. It hurts to think that this is what Jack recalls of his father.

The principal, however, seemed unaware of Jack's diabetes and of any impact the disease might have on his conduct. He assured me that Jack was with him as we spoke, and was now fine.

Dennis did not want me to leave work to go to town to get Jack, and to my regret, I listened to him. So Jack stayed at the school until it was time for the school bus to bring him home.

That evening, Dennis and I were to attend an Agri-Pro dealers' conference and supper in a town forty miles from Hermann. Dennis's parents would babysit. I insisted on waiting to see Jack before we left, but when he arrived, he just seemed cross and we had no time to talk.

The next morning I learned that after my telephone conversation with the principal, Jack had spent the remainder of the day in isolation, locked in a room, *alone.* He could have had convulsions; he could have gone into a coma; he could have died: No one was with him, no one would have known.

I was so angry that I barely stopped to dress before heading into town to confront those responsible. I asked Dennis to come with me, but he said that he would only mess things up because if they got smart with him, he'd kill one of them. He cautioned me to be careful as to what I said, but all I could think of as I drove to the school was that callousness or ignorance could have killed Jack.

I demanded to speak to the superintendent of the school and the principal at the same time: a mistake. Two male chauvinists bonded together against a mother whom they made it clear they thought hysterical. The conclusion of our interview was the declaration that if Jack behaved the same way again, he would be treated the same way.

The next night was Saturday, Drama Awards Night. Jack had been looking forward to this event for weeks. I asked if his suspension covered this. The reply was that it did, *but* that Jack could attend if neither the superintendent or principal saw him.

No one seemed aware of the message being sent: that it was all right to break rules if you weren't caught. Jack was not going to the ceremony on this basis.

When I arrived home, I tried to find out what had precipitated the reaction Jack had suffered. He could not remember if he had eaten lunch the day before or not.

Jack takes his own blood-sugar readings three times a day—morning, noon, and night—on a Glucometer II, an expensive handheld device the size of a Walkman. A single drop of blood is placed on a plastic strip inserted into the machine, which gives a reading. It also has a computerized twenty-seven-day memory. I asked him if he had done a noon reading. Then he remembered that he was about to, but the teacher he'd sassed had stopped him from going to his locker to do so and ordered him to go at once to the cafeteria. After that, he had memory lapses and was very confused as to the sequence of events.

When I read the Glucometer's memory, my hair stood on end. Anything below seventy is dangerous. For the past two weeks, Jack's noon readings had ranged between fifty and thirty, or lower.

"Why did you let this go on?" I demanded.

"I was going to eat right away; it was okay," Jack replied.

I telephoned Doctor J., whose reaction was as horrified as mine. He reproached Jack for not keeping him informed so that an adjustment could be made to control the situation. He ordered Jack to cut back his morning insulin dose by several units and asked that we call him on his beeper if Jack had another such reaction. He also offered to call the school.

I had supposed that they would listen to a doctor even if they ignored a mother, but Doctor J. reported that the school officials insisted that Jack had simply had a temper tantrum to which their response had been appropriate.

During the three days of Jack's suspension, several tests were given that Jack was not allowed to make up. Consequently, his grades dropped and he was no longer on the honor roll; he forfeited the opportunity to join the National Honor Society in his junior year. His grade point average suffered and we feared this would affect his chances for admission to college.

But the worst effect was on Jack's attitude. Outwardly, he

tried to seem indifferent, but inwardly I knew he was furious and indignant. I made contact with a counselor affiliated with the American Diabetes Association and eventually Jack came to realize that the ignorance and pettiness of some of our local officials was far from universal. The result has been an increase in Jack's determination to succeed at any cost, to prove to the school authorities how wrong they were.

Several weeks later Jack learned of an opportunity to go as an exchange student in June to Germany. Hermann is a sister city to Aerolsen, where a host family would board the students. Despite our usual shortage of funds, I wanted Jack to have this chance. I really thought he needed it. I knew it would take some doing, but I was determined to find a way.

# Chapter Twelve

After a severe rainstorm, the Missouri and Gasconade rivers were reported to be rising fast. The Corps of Engineers had opened the floodgates on the Bagnell Dam at Lake of the Ozarks in the southwestern part of Missouri. Dennis and I drove to town to pick up some emergency grocery supplies and found that the creek in front of the high school had flooded the highway. Frene Creek, which begins on our property, commonly floods after a heavy storm; in this kind of weather the high school, which is located near the creek, is let out early so the students aren't stranded. When we passed the school there was already three feet of water covering the football field and a stalled car blocking part of the highway that the police and a wrecker were trying to haul to higher ground. Nevertheless, as we were already so close to town, we decided to go on.

The flood marker by the side of the road was already invisible. Just as we passed the school, the fan belt sucked water into the motor and our full-size four-wheel-drive Ford pickup died and had to be pushed off the road by a tractor. The police closed the road behind us but we finally made it to town. We bought our groceries and drove back by way of the hills.

The next evening we found the waters had risen to within a foot and a half of the twenty-foot levee that protects one section of our property. For fear the waters would rise still higher and destroy the levee or burst through it, Dennis opened the floodgate and our farmland was inundated for the better part of a week. Two days later, the Missouri River crested at Hermann at over thirty-three feet: Official flood level is twenty-one feet.

Fortunately, we did not have many of our crops in the ground and there was still some hope that we would be able to plant when the fields dried out. Dennis had put sixty-odd acres of corn in the week before; this was now drowning. He wanted to put twenty acres of our valley into corn now. We would have less than ninety acres planted when we had hoped for three hundred. And getting in that last twenty near our home was to be a struggle.

Earlier, the local John Deere dealer had offered to take a trade-in on our 4040 John Deere tractor and to sell us a brand-new 4455 with four-wheel drive. I knew the new 4455 would be a sound investment, but I didn't want to trade-in *my* tractor, Gentle Ben. I knew one day my good old tractor would break down, but I couldn't let it go. It had too many associations for me.

"Give him the 4320," I suggested.

Dennis grinned. He knew all too well how much I hated Goliath. "Can't, Mim," he said. "It's older than the 4040. There isn't enough trade-in value left."

We also owned a 6060 (1982 model) Allis Chalmers, a really nice little tractor. Dennis used it for everything, from grinding feed to planting and spraying. We all considered it his personal property. He had let me use it the prior two falls to sow wheat and I knew it handled well except for the fact that the shift was on the floor beneath the steering wheel so it was almost impossible for me to adjust it. Sometimes I had been forced to sit there with my foot holding down the clutch so long my leg cramped before Dennis rescued me. I never thought Dennis would offer *his* tractor in trade.

He surprised me and the dealer offered only five thousand dollars less than he had for Ben, so the deal was made. It was Kevin who cried when they came for the 6060, and each time we passed the John Deere dealership on the highway for several weeks afterwards, he looked for "Daddy's tractor."

We didn't have the new 4455 in the valley when the floods cut it off, so we had to work without a four-wheel drive. And now Ben had a huge hole in the muffler so that it ran rough and noisy, and black smoke belched over the roof of the cab. Of course a mere hole in the muffler wasn't going to stop Ben but it sure made everything smell bad and it sounded terrible.

The ground was so saturated after weeks of rain that almost at once the disc got stuck. Dennis lifted the disc blades from the mud and attempted to free it by shifting gears to rock the disc back and forth. That didn't dislodge it. Then he hooked one end of a logging chain to the frame of the pickup and put the other end around the front axle of the 4040. The truck and tractor were on solid enough ground to get some traction and finally the disc was freed.

We had to avoid that low section but we finished discing and hooked on the corn planter. Then it started to drizzle. But Dennis was determined to keep going. I have never seen him plant so fast. I wondered if the planter could roll the seed out that quickly and whether Ben, belching smoke even faster, would hold up. Even so, all he could finish planting was about ten acres and three weeks passed before we could do more as the rains and flood persisted.

It was June 2 before we could start to plant again. And we were forced to leave patches of low-lying ground—"swags"—unseeded because they were simply too wet to plant. (Some swags retained several inches of water all through that summer.) School had let out so the boys were available to help.

Little Kevin spent his days opening the twenty- and fifty-pound bags of fertilizer and seed to be dumped into the planter and fetching and greasing—the tasks I had started with. Jack continued to care for the hogs every day but often came out to the field to disc so we could keep two tractors working even if I had to be elsewhere. He even liked the power of Goliath, our 4320. But he is a true redhead, and a day on the tractor crisped him. And then he had problems with his blood sugar. Sitting on the tractor wasn't enough exercise for him and so his blood sugar rose to a dangerous level. He had to do something really energetic to get his count back down to normal. He joked that working in a long one-hundred-twenty-acre field, at the slow pace of a tractor pulling a disc, together with the warmth of the sun, the bouncy motion of Goliath's broken seat, and the hum of the motor, put him to sleep.

This was all Dennis needed to hear: He fixed the seat in a hurry.


Dennis has an impatient, bossy, perfectionist, Type-A personality. The mere thought of failure is enough to drive him to the edge. He is determined not only not to fail, but to succeed and succeed big. The problems we continually encountered with the machines, the weather, and our health wore on all of us but especially affected Dennis.

Twelve- and fourteen-hour days were hard enough for us, but Virgil was getting older and the demands made upon him were making him fatigued and irritable too, and defensive. I have to give my children's needs top priority, and I cannot abide an untidy home. When I spend all my time working out of doors, the sight of dirty dishes, unwashed laundry piles, and unswept floors still makes me feel like a failure. I had also committed myself to teaching morning Bible school for one week in the summer, not expecting that we would still be planting in July. Dennis couldn't understand why I wouldn't break this commitment. And then, although he didn't oppose swimming lessons for Kevin, he complained because the lesson times interfered with his schedule. Finally, we reached the point at which we couldn't even speak civilly to one another.

I wrote in my journal:

This farm is killing us! I have watched it destroy my husband's health and our closeness. We argue more now than we ever did and over things of no importance.

We met ten years ago today. On that day I told him I was a "country girl," never dreaming that those words would one day be a source for his ridicule.

I was driving Kevin to swimming lessons and the car had a flat tire right in front of the swimming pool. The service attendant was so horrified by the condition of the tires that he insisted on mounting the new ones Dennis had ordered but had found no time to have installed. So I was two hours late getting out to the field. Virgil gave me the cold shoulder the rest of the week and I had a mixed reception from Dennis. By his standards I am always late, no matter what else I have to attend to.

I didn't even cook anymore. Jack had done so much helping out in the kitchen in the past weeks that I didn't even remember where in the cabinets some of the food items went. I was spending eight to twelve hours doing fieldwork every single day.

Sunday was the first full day I had taken off in a month. Jack was sixteen and he wanted to have some friends over for a party. The kids had a ball. It was worth it, no matter what Dennis said. Yet no matter how hard I tried, he complained that I was never around when he and Virgil needed me.

I was not going to tolerate any more of this! I vowed.

One day I climbed aboard Goliath at twilight, so furious I was thinking of leaving Dennis Fricke and his farm. As the rain began to fall, cold wet drops mixed with my hot tears.

Quitting was not my way. But Dennis had accused me of not caring once too often. I had changed my life to please him, I told myself. I gave up plans to go to college. I gave up a job to care for our son. And then, instead of returning to the work force so I could at least draw Social Security when I turned sixty-five, I opted to remain on the farm to save us the cost of hired labor. Was staying in the fields every night until nearly midnight evidence of my lack of concern?

The drizzle got heavier and Dennis decided we had to get the grain truck from a wheat field several miles away where Virgil was combining. We had to put the pickup, loaded with bags of seed, in shelter at Ted's shed first. We worked in silence: I refused to speak to Dennis and he acted as if he had no idea of what was bothering me. By the time we went home to a supper of canned soup at midnight my temper had cooled. There was no use trying to reason with Dennis, but I was not going to quit either.

The next day we returned to the field and were confronted by nothing but breakdowns. A bearing went out and then an axle warped on one disc. I spent that evening trying to pull a disc that was breaking apart from the main frame. It had to be welded over and over but continued to give trouble.

Dennis seemed to react in his usual cold fashion, but then, when he thought I wasn't listening, I heard him tell Virgil, "I can't handle any more of this. If we ever have a flood in the spring again, I'll declare bankruptcy before I try to plant the ground!"

I was stunned. Then I realized that we had coped with frustrations before, but now it seemed the land itself was against us. We were all near to losing our minds because of it, and taking it out on one another.

We reached the final seventy acres to be planted at the end of the first week in July. The next morning we drove Jack sixty miles to the special camp for diabetic children he attended annually, and were back in the fields again by three that afternoon. I was so tired that when I slept I didn't really rest; I awoke feeling that I hadn't slept at all, only to spend the day running at full speed. I was sunburned again, this time on my legs, because I had been wearing shorts while driving Goliath, which has no cab. Kevin and I got poison ivy. I developed diarrhea. And Dennis's blood pressure soared. I could not convince him that mixing beer with his strong medication was not helping him. He was sure the beer relaxed him. I think it rendered him numb.

The final piece of land had stood in water for some days. Now it was dry, but littered with residue from the river and it would not break up evenly. The sun's heat had baked the soil to a hard crust, but a foot below the surface was gooey mud. When we disced the wetter soil up to let it set before planting, it turned into gritty clumps that flew about, coating everything or bunching into mounds behind the discs. Virgil was driving the new 4455—now named Godzuki, pulling the disc I called Godzilla—and I was once again on the monster tractor, Goliath.

I was so concerned with just getting from one end of the field to the other that I paid no attention to what was going on around or behind me. So I didn't realize that I had lost the bolt that attaches the hydraulic cylinder. But I heard the *pop* when the cylinder broke loose from its casing. The third disc had broken. I could use it but I couldn't lift it when I turned. I could have managed, but the clumpy earth packed beneath the tongue of the disc so that I was pushing a mound of dirt instead of opening up the soil in back of the tractor. The only help for it was to stop, back up, and drive around, to attack from a new direction.

And that's how I finally learned to back up a tractor while it's pulling a four-wheeled object. Dennis thinks this is the silliest thing he has ever heard called an accomplishment. But I saw the grin on his face when he watched me tackle those mounds of dirt and conquer them.

.   .   .

Jack left for camp, and Kevin and I took over his chores in the feeding buildings. Every day we watered and fed and ran the auger in the fattening building. Dennis was spraying the crops, a task he can handle alone. No sows were farrowing. Without Jack we were shorthanded, but it looked like we would be able to manage pretty well although we had three hundred weaned piglets in the farrowing room.

We don't leave them in there for long because once they discover they can jump over the nearly two-foot-high walls of their pens they get into everything. A hose forgotten and left on the floor for a few hours will be bitten full of tiny teeth holes. Buckets and loose feeders will be nosed around and rolled until they crack and fall apart. Wire is torn loose, drinkers are disconnected, hoses are mysteriously shortened. Anything made of paper, cloth, or plastic simply disappears.

Two years earlier, a sow had broken out of her crate in the middle of the night and then into the middle room of the farrow house where the feed is stored. (It took some ingenuity to get her back into her pen without injury after she charged Dennis.) Now our piglets discovered that the door to the middle room had a little give where the sow had torn the metal partition away from the frame and they nosed at it and chewed it until they broke the whole lower section of the door loose. Then all three hundred of them must have charged into the middle room at once.

Dennis buys fat as a supplement to feed to sows after they

farrow while they are nursing and to the young pigs so they will gain weight. It comes in fifty-pound bags. That night there were three bags lying on a pallet on the floor. The pigs tore those bags apart and distributed the contents all over the room. When Dennis stepped into the middle room the next morning he nearly fell flat on his face. The floor was covered an inch thick, wall to wall and up the walls, in grease. He tried to hose it away, but it was almost pure fat, and cold water wouldn't budge it. This hosing just made it congeal and become even slipperier.

Kevin and I carried some hot water down from the house, but by the time we reached the middle room it was too cool to be of help. We tried mops saturated in degreaser, Mr. Clean, and ammonia but we just smeared the fat around. Finally, we resorted to scraping the stuff off the floor and walls with shovels and piling the residue in the middle of the room. After several hours of labor, the floor remained shiny and slippery but it was possible to walk on it with care.

That night the pigs broke into the middle room again. So Kevin and I spent the next afternoon scraping up the mess all over again. Then we scooped it up and shoveled it into the lagoon.

We were all glad when Jack returned home. I was more than glad; I was overjoyed.

Jack brought home from camp every piece of clothing he owned, dirty. I did a huge wash and hung it on the line, then went to help Dennis move the tractor and spray tank, telling Jack to get his clothes in if it began to rain or he'd have to go naked the next day. It began to rain and Jack ran to get the clothes in, thrust his feet into his boots—and was stung by a scorpion on the arch of his foot.

Our Missouri scorpions are only about an inch long and their bite isn't venomous enough to be harmful to most people. But it was far more dangerous for Jack, who seemed to be allergic to scorpion stings. His foot swelled to nearly twice its size and his leg turned blue to midcalf. We drove him to the nearest hospital emergency room, luckily only eight minutes away. This small facility is a way station for the better-equipped hospitals in Columbia and St. Louis with which it is affiliated, so his doctor was able to prescribe by telephone an antihistamine injection to be given upon our arrival and it worked. Now we had to watch out for a severe insulin reaction, but luckily Jack escaped without ill effect.

# Chapter Thirteen

As a rule we wean our pigs when they are five weeks old. In the larger farrow house we run the sows out and leave the piglets in the pens for another week or two before moving them to the finishing building. We also have a smaller farrow house that has a nursery, but when we had to reduce our herd from two hundred and fifty to less than one hundred sows, we no longer had need for both buildings.

Now Dennis decided to use the small nursery again. It had

stood vacant for more than a year, and the dirt that had accumulated in that time amounted to a thick, dusty layer. Although the water pipes had frozen here too, a hose could be strung to carry water to the pens, and a little hosing with fresh water got the interior sparkling clean. One Saturday afternoon Jack, Kevin, and I moved about sixty little pigs from the farrowing room of the smaller building into the adjoining nursery—well, Jack moved most of them. We cleaned the feeders and filled them with fresh ground feed and were delighted with our achievement.

When we returned the next day to feed them we discovered several dead pigs. This happens sometimes when piglets are moved. The stress of being carried from one room to another might have been too much for them, we surmised.

But within twenty-four hours a dozen pigs died. Dennis went to feed them and reported that the pigs were coughing so hard they couldn't walk. When I went to look, what Dennis called coughing looked like seizures to me. The poor little things were choking on saliva that foamed from their mouths, and their bodies jerked so hard they no longer had the strength to stand. Then they lay in the throes of death for some time before actually dying. I couldn't bear such suffering. And I couldn't understand why the ones that were dying were those Kevin and I had moved; most of Jack's were all right. What could we have done to them?

We transferred the sickest piglets into one pen. Then we asked ourselves if there might be something toxic in the nursery. It seemed unlikely that just one section would be contaminated when the entire room stood over an open sewage pit and the pens were separated only by three-foot wire rails over a wire floor.

By the third day, when more than two dozen pigs had died, Dennis called our feed dealer. Pete came to examine the hogs and consulted the veterinarian affiliated with his dealership. The conclusion was that the pigs that were dying had an intestinal disorder caused by too much protein. They liked their new feed so well they had eaten too much, too fast. We lost nearly half of this group of piglets but it was just a coincidence that the pigs Kevin and I had moved had been the hungriest ones and so suffered the worst.


Occasionally sows farrow unexpectedly in the pens out of doors. This can be a disaster as it usually happens when it is too cold, too hot, or too rainy and the piglets do not survive. The sow makes herself a nest—a hole in the ground she can wallow in comfortably—but if it rains, this hole becomes a mud pond and the piglets drown in it. Of course, no matter where she farrows, a sow may accidentally lie on her offspring, killing them. And some will deliberately kill and eat their own litters. If there are other sows in the pen they may trample and devour the piglets. Or a sow may have a difficult labor as she lies undetected, hidden by other sows, and in the course of a night or even a few hours die with her unborn piglets. But these mishaps are more likely to occur when the sows are out of doors, unsupervised. So we try to have all our sows deliver in the controlled, stable environment of the farrowing rooms.

As a rule, sows that have farrowed before are not hard to move in or out of buildings. But when they are left out of doors until they are close to delivering, they want to stay in their nests and it's nearly impossible to move them. We try to keep the sows that are close to term in pens inside the finishing building where we can watch them closely and move them more readily, by truck, when the time comes, over to the farrowing room. But the inexperienced gilts still cause problems. They are only used to being in wide-open pens. When they find themselves in the confines of a dimly lit building they become frightened, and when frightened, they run. So whenever we can, we put a few seasoned sows in with the gilts to teach them and calm them.

Of course, to get the right animals to the right place at the right time takes a certain skill in judgment. I have picked a sow whose belly was nearly dragging on the ground only to have Dennis tell me she wouldn't farrow for several more weeks and be proved right. If a sow is producing milk, she will farrow "soon"—from within twenty-four hours to a few days. Otherwise, it's still a guessing game to me.

Labor Day weekend we attended a party at the home of one of our landlords. Jeff and his kids were there. Bonnie's absence was palpable.

Dennis went as a delegate to the annual Missouri Farmers' Association convention in Branson, Missouri. It was the first time he had left home without me since our marriage. I had been in the hospital several times, but he had never left *me*.

I assured him I could manage for twenty-four hours without him. The more confident I sounded, the more apprehensive he seemed. What would I do in an emergency? He was taking the car and he knew I'd refuse to drive the pickup.

I reminded him that Virgil was right across the valley, that Ted, or Pete, or any of our other friends could be appealed to.

I must have seemed too calm. Dennis accused me of not missing him at all.

But I did. I missed him, but I had no trouble coping without him. Maybe that's what upset him so.


Dennis and I spent the fall in the fields. Combining the crops ran into late November, and after Dennis combined the crop, I chisel-plowed. Like discing, chisel-plowing is easy if you pay careful attention to what you are doing. But most often I ended up doing this task late in the afternoon or in the evening, and I don't like working in the fields after dark.

Dennis and Virgil love the fact that the night closes in and you feel totally isolated, alone on the tractor. That is precisely the thing I do not like. Even on a clear, starry, moonlit night when objects are as visible as in daylight, you cannot see beyond the tractor's light. You cannot see where in the field you are. I like to look all around as I work, to see the hills, the line of trees that edge the woods bounding the far side of the field, the deep ditch at the corner. When I'm on the tractor at night I cannot suppress the sensation that someone or something big is about to jump out at me.

Most years Dennis tries to be out of the fields by the third week in November as a precaution. Hunters occupy the deer stands that dot the woods and stay overnight at the otherwise vacant houses on the properties we farm. The noise from our vehicles would spook the deer, and of course, a shot into the fields might strike our machinery—or us. But we were late this year and had to take our chances.

Earlier in the fall in the first field of soybeans Dennis combined we had found the remains of a young deer among the bean plants. It had struggled, trying to get up, knocking down soybean plants and pawing the earth, before it died. Someone had killed it out of season, someone who tracked the deer from the road, never venturing from his vehicle, who'd startled the little deer with a high-beam lantern as it fed at night. This little deer had not been killed by a *hunter.* But hunters often take the blame for these acts.

Until hunting season began, as we worked outside at night, we did see a few "spotlighters." They drive around shining their lights into fields to see if the deer are feeding, but their disturbance of the animals is brief. And as they don't shoot, we don't resent them.

When the season started, Dennis shot his deer at Grandpa Hugo's farm, an old scarred buck whose antlers had been broken in past fights.

. . .

In December Dennis wanted to chisel-plow a section of land that still retained stumps beneath the surface soil. This land had been cleared twenty years earlier, but the oak stumps and roots deep in the ground were slow to rot. We had avoided mallboard-plowing this section, since this type of plow goes much deeper, and striking a large root would severely damage it. But the spikes on a chisel plow gnaw the dirt around the roots and help raise them to the surface.

It was my job to follow behind the tractor to pick up the pieces of root and stump the chisel plow brought up. Some were so big I had to leave them for Dennis to haul. I started getting a rash on my thighs where I had to brace the larger root limbs to carry them. I thought it might result from an allergy to dirt or from the chemical residues of the herbicides that we use that remain in the dirt, but then I started getting a skin rash from feeding dust in the hog buildings too. In a strange way, that was a relief. No rash is fun, but I didn't like the idea of having a reaction to chemicals. We are very conscious of the danger of chemicals and alert to their possible misuse here on the farm.

Farmers must obtain a permit and the permit must be renewed every year for any substantial chemical use. If we are found to have violated the safety rules for chemical use, our permit will be withdrawn. Regulation for safety's sake is essential because it is simply fact that we *must* use chemicals in the course of our operations. Before the Civil War they used field slaves to laboriously chop weeds and pick the insects from plants on large farms; during the Great Depression, when farms were reduced to thirty or forty acres, a farmer could chop his own weeds by hand. And if the grasshoppers came along and ate the crop, there was nothing he could do about it. We cannot operate in the old way. We are dependent on chemicals to keep down weeds and insects.

As we have expanded, Dennis has started having the land custom-sprayed each spring. The sprayer truck operators are professionals and the result should be more efficient and safer application of pesticides and herbicides.

Corn and soybean plants are sprayed when they are a few weeks old to preserve them from insects and weeds. The chemicals used are species-specific. Cockleburs—which choke out everything—and Johnson grass are more difficult to eradicate than woodbine and beautiful morning glory, which forms a death trap of tangled vines in soybean fields. At harvest we often find late-growing weeds but at least the corn and soybeans have become strong enough to outnumber them.

Wheat is sown in the fall, so insecticides and herbicides aren't needed to protect it. The only chemical we apply to wheat is fertilizer that contains nitrogen, phosphorous, and potassium, the same ingredients found in manure. The fertilizers we sometimes use in spring—urea, anhydrous ammonia, and liquid nitrogen—are also naturally occurring chemicals, but they are highly dangerous on contact to eyes and skin and poisonous if ingested. It is the farmer, rather than the ultimate consumer, who is in the most danger from the chemicals we must use. It is for this reason that we treat all dangerous

products with extreme caution—and why we must be licensed
to use them.


Virgil's back had been bothering him, a reminder that he
wasn't getting any younger. Soon he could barely walk and had
to get six weeks of rest before his strained muscles improved.
We had Christmas at our house for Dennis's family. Dennis
went to fetch Virgil, who came in walking on two canes, then
went back to get his mother and all her contributions to the
feast.

Jack went to spend a few days with my parents after
Christmas, and Dennis and Kevin spent the New Year's week-
end viewing *thirteen* hunting and trapping videos on a rented
VCR. I went to my room and put together a crossword puzzle.

New Year's Day my stepdad brought Jack home and my
favorite aunt, Aunt Liz, came along to pay me a visit. The day
was warm so the roads were slushy. But here in Missouri we
say, "If you don't like the weather, just stick around." It thawed
in the daytime and froze at night every day for three weeks
after that. The roads were sheets of ice, and even vehicles with
four-wheel drive were risky to drive.

Jack had been doing most of the wood chopping since
Thanksgiving with Dennis felling the trees and trimming
those felled the previous spring by the woodcutters. Now they
were working in the woods right behind our house and I was
getting cabin fever so I was happy to oblige when Dennis
requested that I come to help them stack wood.

I slid across the driveway that circles the back of the house. It's usually in shade and so retains ice. I made my way uphill by hanging onto the trees I passed. I managed to cross the drainage ditch behind the house and climb over some dead logs, following in my sons' footprints. But when I reached the slope at the bottom of the woods that area was so slick I couldn't stand up. When I let go of one tree trunk to reach for another, I slid onto my butt and could not get back on my feet. I finally had to crawl. I could see and hear Dennis and the boys not a hundred yards away walking from stack to stack easily, but on the slippery slope I couldn't manage to stay on my feet.

Finally, I sat on my rear, pushed off, and slid from tree to tree back the way I had come, back across the drainage ditch, over to the car. I was able to walk to the drive but then it was back on my butt again until I reached the front door. I landed on the kitchen floor, soaking wet clean through.

Twenty minutes later Kevin came looking for me. Dennis wanted to know where I was. It turned out Kevin and Jack had chosen that route to the woods because they wanted to go the slipperiest way so they could slide.

When Dennis came in for lunch, he wouldn't believe that I hadn't been able to make my way out to the woodpile to help them. So he drove me through the hay field and up to the woods where I had to crawl up the slope but, once on top, found walking was easy especially where the ground had already been powdered with sawdust.

We cut and stacked that evening and the next day. The following evening it was pretty cold by sundown, but we kept stacking by the light of the moon, hoping to finish. It was still and quiet, with no sounds other than those we made or the grunts of the hogs in their pens in the nearby clearing. The air was crisp and clean. The reflection of moonlight on the snow furnished ample illumination. But by the time we finished, the temperature was below freezing. Dennis was wearing boots with cleats so he was able to walk, carrying the saw, but for Kevin, Jack, and me it was another butt-sledding trip home.

We survived Jack's cutting himself in the left knee with the chain saw. I had to clean it and dress it; the roads were not passable for several days. We survived ice and snow, and progressed into February. I still had no functioning washer or dryer so Dennis had to drive me to town to do the wash in the evenings. I thought this would provide some impetus to getting me new machines, but it turned out he liked doing it—he enjoyed the peace and quiet of the laundromat, or so he said.

At this time we had fifty sows farrowing in rapid succession so we had to use the extra farrowing building again. Dennis and Kevin took care of them while Jack did the daily chores in the big farrowing house and finishing building. I was working on the books for the accountant and helping with the sorting and selling of the hogs.

One day Dennis and I weaned about a dozen litters and immediately turned the sows into the outdoor pens with the boars. Dennis prefers to allow the sows to recuperate for a while after weaning before turning them in with the boars but we were really pressed for space and had no choice.

Unfortunately, nearly all the sows we moved went down—fell to the ground and would not arise. They came into heat and in their excitement, roughhoused with one another. The larger, older sows climbed on the backs of the smaller ones and rode and injured them. We had to get the injured sows away from the others, or they would never recover. A downed sow will not be left alone by the others; instead, they will worry at her until they kill her.

The injured sows couldn't walk but they could still bite. They were in pain and consequently bad tempered. We finally decided to take two heavy chains and wrap each sow around her middle just behind her front legs and just before her back legs, attach each chain to the endloader on a tractor, lift her, and put her in a holding pen. Virgil carefully worked the lever that controls the hydraulic cylinder that lifts the enloader. And slowly, with a few mishaps along the way, and a constant accompaniment of frenzied squeals, we moved ten sows. Three were on their feet again in a few days. Four took a few weeks, but made a full recovery. One died that day, and two had to be shot.

On February 22 we finally bought a matching Maytag washer and dryer, the largest the company makes. It cost us one thousand dollars, but I think it was a good investment when you consider what I was spending on trips to the laundromat.

When they were delivered, I spent the next five days washing anything and everything I could get my hands on.

Here on the farm we had braved the deaths of dear ones, floods and hardships, but our homes had not been invaded, our freedom and privacy were not threatened by the army of a foreign country. I could not begin to comprehend the terror and devastation that the people of Kuwait must have felt. My heart went out to them and to our men and women who had gone to the Mideast on their behalf and on behalf of our country.

Our problems, in comparison, seemed minor but again, nothing to do with your children really seems minor to a parent. Kevin, in second grade, started refusing to go to school.

It seemed that one evening, while I was driving Jack to a meeting in preparation for Confirmation classes, Dennis had helped Kevin with his spelling homework but forgotten to sign the homework paper. The next day, instead of telling the truth, Kevin signed the paper with my name and told his teacher that I had been so busy he had to write it for me. When she didn't reprimand him, Kevin got upset because he had told a lie and gotten away with it!

He came home crying and from then on it went downhill. I went to the school to talk to the principal and his teacher, but I still had to put Kevin on the bus by force several mornings in a row. The school sent Kevin for counseling, but unknown to them, he liked the counselor so well that he skipped physical education to go to see her again.

Then Kevin began to cry for his father every morning when it was time to go to the bus stop. Dennis started to make a special effort to see Kevin each morning and to talk with him before he left for school. But at the same time, Dennis told Kevin that he had better straighten up quick: Each time he lied to his teacher, skipped school, or made a fuss about getting on the school bus, he would be docked one week of television watching. That ended that. And by the end of first semester, Kevin was getting straight A's.

Spring opened with sixty-mile-an-hour winds that blew down trees and sheds and lifted off roofs all over central Missouri. We lost a part of our house's tin roof, but before we could repair it, it was time to plant corn. We made a start and then the rains came, so we had to start again two weeks later. But though the river rose, it did not flood this year.

Dennis had a filling in a tooth repaired, and after that he began to suffer excruciating pain in his face, in addition to pain in his ears. His blood pressure jumped up. The dentist sent him to a specialist who performed three root canals, and voilà! Overnight the pain subsided and his blood pressure began to go down. It was like our own miracle.

I was discing, pulling Godzilla, near Paula's house when I popped in to pay her a visit.

"Bonnie died just one year ago today," she told me. I was overwhelmed with guilt. I had forgotten.

I realized that on May 1 I would turn forty, which brought on depressing thoughts of my own mortality and my lack of accomplishment. I remembered that when my mother had her fortieth birthday, she told everyone, "I've already lived half of my life." She seemed to feel that she had failed in those first forty years and could never make up for her mistakes.

"Over the hill" was the phrase that kept going through my mind. I started waking up at night in a cold sweat.

Kevin, who has a summer birthday, had chosen mine as the day to celebrate in school with his classmates. I knew this, but I was so preoccupied that I forgot. I had nothing baked for his class party and had to call the bakery in Hermann. Luckily, they were able to bail me out, and his classmates never knew.

Eventually, I concluded that there is no such thing as "over the hill." Life is all uphill, one step after the other, as we pursue our goals. Each decade brings us closer to the top and, inevitably, to death and our reward beyond death. The closer we come to the top of the mountain, the thinner the air gets and the more difficult the climb. We have to learn to pace ourselves to get there. I see and hear of so many people who accomplish things at eighty that they would never even have aspired to at forty that I refuse to adopt my mother's negative outlook. Who is to say where the midpoint of one's life is?

I have forty years of achievement and I am proud of the blood, sweat, and tears I've shed to get this far.

# Chapter 14

Share the Fun Nite brought our 4H Club together once more to write a skit and rehearse it. This year I was sure we had a winner. Jack did most of the scriptwriting, and our theme was environmental awareness: We have to fight pollutants every day if we want to save the land. I think we gave a terrific performance, but to our disappointment, our club didn't even get an honorable mention.

Jack had a date for his prom with one of Bonnie and Paula's

nieces. We let him take the T-Bird. But getting him ready and into his rented tuxedo was a real labor. And it has been so long since Dennis has worn a tie that I had forgotten how to tie one. I had to telephone Dennis's mother to get step-by-step instructions. Then Jack left the house wearing his best sport shoes because he refused to drive in the dress shoes I had bought for him. But I guess he had a good time. And three weeks later he was confirmed in a solemn ceremony at St. George's Church, along with thirty-one classmates.

I had a new test of my abilities as a farmer. Virgil was filling ditches and building terraces with the endloader in one of the fields that had washed out quite deeply in our recent storms, so when Dennis decided the moment had come to castrate some hogs, he assigned me the job of cutting them while he held them.

I tried it. Once. But the feel of the pig's soft skin slipping beneath the sharp blade of the knife made my stomach churn. I could not press down with the knife hard enough to cut the skin. My actions were so slow and tentative that the pig had plenty of time to squeal and to wiggle in vigorous protest. Then I couldn't find his testicles to snip them off. Dennis was so aggravated, but I had never held myself out as a surgeon. The lesson I learned was that this farm-wife stuff must have its limits. After that one try, I held the hog while Dennis did the cutting—zip, zip—in his usual manner and plopped the severed testicles onto the floor to be hosed away later. (We have not saved them for a mountain oyster party in some time.)

. . .

By the first of June we were much further ahead with our fieldwork than we had been in 1990. That year by this date we had planted just a scant ninety acres of corn and no beans. This year, 1991, we had two hundred acres of corn and beans in the ground although some of the crops were slow and yellow looking because they stood in exceptionally wet soil. Nevertheless, we could maintain a sense of humor despite our problems because our equipment was new or well-repaired and Mother Nature no longer seemed quite as determined to do us in.

Saturday, June 1: Dennis was up at six-thirty and I woke the boys at seven. Neither was happy about this. Jack had attended the high school commencement followed by a pizza party with his friends the night before and had not come in until midnight; Kevin always wants to sleep late. Jack took his insulin shot and grabbed a bite to eat, a routine he has to follow first thing every morning. By seven-thirty the three of us were down in the garden. It was a cool morning, sunny but breezy, perfect for working out of doors. Jack began to hoe the tomatoes and beans, one row each; Kevin started with the rows of beets and broccoli. I went to work cutting off the spinach because it was going to seed, and since some heads of broccoli were so large they had begun to flower I cut them as well. The cauliflower was in need of help: I babied it a little, tieing the leaves over heads that were just beginning to form, but I had small hope for it. The weather had been too wet and too warm, and grasshoppers had already attacked the leaves. But the cabbages were doing so well that we powdered them to keep the bugs away.

When Jack read the label on the package of Sevin that we were using for the cabbage, he discovered that it could be used to detick and deflea dogs as well. He just had to try it on Wiener, who loved the attention even though the white powder made him sneeze. Sally made herself inconspicuous, lying down on the edge of the wheat field that borders the garden where she continued to eye Wiener and Jack with suspicion.

With Wiener supervising, the three of us attacked the weeds between the rows of onions. Although the handle of a hoe broke, we kept going. By eight-thirty we were ready to return to the house satisfied with a job well done.

While we were in the garden, Dennis had done his morning check-up chores and driven into town for a wagon of loose fertilizer and some containers of chemical spray. Jack headed for the hog house to water some just-weaned pigs while Kevin and I had our breakfasts of orange juice and fried egg sandwiches and Kevin sat down to watch Saturday morning cartoons. Then the boys cleaned their rooms while I did three loads of laundry and hung it out to dry, plus a few more indoor chores.

When Dennis returned, Jack took the car to go to town to get a haircut. Dennis wanted our help in moving some sows out of the farrow house, so Kevin and I set out to walk around the ten acres of wheat and through the fescue on the valley floor to reach it. The fescue was three feet high, but soggy and lying down, beaten flat by last week's storm. Walking was not easy due to the hazard of muddy puddles hidden beneath the layer of fescue. But that hazard diminished in importance when I gripped the electric fence surrounding the pastureland preparing to cross it, and found that the wire was live. Kevin had gotten far ahead of me, but my scream brought him back, on the run.

The charge is not enough to really injure anyone although it's more than a tickle. So I had to detour through the lower end of the wheat field and go around the machine shed to enter the hog buildings area by the front gate.

By the time I arrived, Kevin had already helped Dennis move the first four sows through an aisle in the nursery and into the narrow twenty-foot passage between the farrow house and the finishing building through an improvised aisle between a fencing panel and Dennis's pickup truck. I was impressed into service to hold up an eight-foot-long piece of tin to block the sows should they try to go anywhere other than through the door of the finishing building. Soon Dennis became too impatient to move any more sows because he was determined to plant corn.

He loaded ten bags of fertilizer onto the truck to be deposited with the seed, and since he also had to carry a spare tire, spray chemicals, toolboxes, and the portable fuel tank, there was barely room for Kevin's bicycle and none for the cooler, both of which were deemed essential. I was delegated to bring out the cooler in the trunk of the T-Bird when Jack brought it back from town, after first driving to town to buy some apples and some ice.

By the time I caught up with them in a field Virgil had started discing earlier that morning I was instructed to take Kevin with me to the storage shed in another section to fetch two more bags of seed corn. When we had completed this errand, the field had been disced once and Dennis had just spread the loose fertilizer—quick work for a twenty-five-acre field. But the sky was overcast and he had to work fast.

We quickly filled the planter with seed and fertilizer and Dennis began to plant. Kevin and I cleaned out some buckets in the creek and cleared some rocks from the field that had been turned up by the discing. By four o'clock the field was planted. Then we unhooked the corn planter and attached the five-hundred-gallon Kuker sprayer. Dennis, Kevin, and I squeezed into Ben's cab and rode around a group of farm buildings and across another field to a deep part of the creek to fill the sprayer tank with water.

First I tried standing on a rock and dipping up the water to try and save my shoes. But the rock was slippery and after a short time, disgusted and dizzy from bending all the way over and straightening up rapidly, I took off my shoes (and tossed them in Dennis's direction) and got into the water. It was barely calf deep but so cold I could feel the icy shock through my bones right up to the small of my back. I was sure I didn't want to end up sitting in it.

The tank already contained seventy-five gallons; we had to add another fifty using three five-gallon buckets one of which leaked. I stood barefoot in the water, filling each bucket, then handing it to Kevin who pushed it to his father standing on the sprayer tank step, emptying each bucket into the tank. When the tank was full, Kevin wanted to take a swim, but one mouthful of green algae and he was ready to get back to work.

By seven that night we were ready to move the tractor and spray tank to the next field, six miles away. Then we checked another field: still too wet to plant was the verdict. And then home to hot baths and a supper of canned soup and pizza made from a Chef Boyardee mix.

After Mass on Sunday, I began to disc while Dennis sprayed a nearby field of already planted corn. I disced thirty acres in four hours and was very pleased with myself.

The next day I helped Dennis move the drill so he could start planting soybeans. Virgil arrived to begin the second discing of the field I had started, with Dennis sowing two rows behind him. School had been extended into June to make up for the time lost because of snow and ice in January so Jack had to study for final exams after he did his hog chores. In the early evening hours I disced and Virgil sowed, while Dennis and Kevin went to get more soybean seed. Three good days, full of accomplishment.

Of course, it rained the next day. We could get nothing done and had to race to get the load of seed under cover before it got soaked and ruined. But that evening our spirits picked up when we found that although it had rained all day, not more than seven-tenths of an inch had fallen, so we could plant again in a day or two.

I took Jack into town to purchase traveler's checks. His diabetes was now well controlled again after weeks of almost daily insulin and diet adjustments, and Doc J. had given his approval and prepared prescriptions and letters of explanation for Jack to take with him. His teacher-chaperone and several classmates had been educated about diabetes and were carrying sugar, just in case. And his German host parents were physicians. So Jack *was* going to make the trip to Germany, using money we were allowing him to withdraw from his college savings account that he had earned by selling hogs at the fair. I think I was almost as excited as he was about the trip.

When I returned, we moved the sprayer tank to the smaller hog house where it would be easier to fill the next time. I was not anxious to get back into that creek again. Then, into the fields, which, to Dennis's delight, were drying rapidly. He decided to sow the few rounds that remained right then and there.

I was to turn the 4455 around and to disc past the drill exactly one foot beyond the last row that had been sown. It should have been easy. But it was late and I was careless. I forgot that the harrow on the back of the disc stuck out a foot wider than the final disc blade. As I passed the new drill and began to throttle down, the harrow on Dennis's three-month-old disc struck the wheel of the drill. Kevin was almost knocked over, and the wheel on the drill was bent, the tire gouged, and the harrow kinked.

"How in hell did you do that, Mim?" Dennis asked indignantly. "A hundred-acre field and you manage to hit the only thing in sight—my new drill with my brand-new disc!"

I felt absolutely terrible about it, and this time I had no excuse.

The next day we started with sorting sows, then I disced for an hour or so, and made a quick trip to town. On my return, I helped Jack begin to pack. In the late afternoon I was fifteen miles away, pulling the disc through the wide-open ground that bordered the green, tree-covered banks of the rushing Gasconade River. Two hours later, we had finished planting this section and returned home for another canned soup supper. The next day, more of the same, plus sorting leftover seed to see if it was plantable. Alas, most of it was moldy.

That afternoon I did a major clothes wash, the dishes, helped Kevin rearrange his room, and checked Jack's suitcases with him one last time. Paula stopped in for a visit, bringing three dozen fresh eggs from her grandmother's hens. Then we went to meet Dennis in the fields.

Goliath (our 4320 tractor) had been hooked up to a battery charger yet its batteries kept dying. We went to check while we were waiting for Dennis and found that the cows in the pasture had stepped on and unplugged the charger from the outlet and then nearly chewed the electric cord in two. We turned it off, unplugged it, and put the cord and battery charger safely inside a building.

In another field, Dennis was dumping rainwater from the tarp covering the wagon load of bulk fertilizer for corn. When Kevin and I returned, we moved the wagon to a field that we had had commercially sprayed that day while Virgil was at work discing the spray under. Not much later, Virgil struck something hidden in the weeds at the edge of the field and the harrow was bent at a ninety-degree angle on one side and needed on-site repair.

It was dark before we quit work.

Jack left for the airport in St. Louis. From there he would fly to Chicago, where he and thirty-two other students would board a Lufthansa flight to Düsseldorf. Then they would take a train trip to Paris, tour that city for two days, and finally travel on to Aerolsen.

Kevin's first Communion class was recognized at the ten-thirty Mass on Sunday. The girls wore frilly white dresses; the boys, dark pants, white shirts, and ties. After the sermon in their honor there was a small reception.

On the way home we stopped for a few groceries. And then I wrecked the T-Bird.

Traffic was heavy. The woman driving in front of me braked but did not signal for a turn until too late. The headlight of the Thunderbird got hung up on her taillight. Seven hundred dollars' damage to the left front panel and left front fender. Boy, did I dread telling Dennis.

His day had gone very well until I failed to show up by two-thirty and he realized something was wrong.

When he finally took a look at the car that evening, he only shrugged and said, "I hope the insurance company doesn't decide to total the car." He even took the time to pry open the left headlight cover.

We spent half of the next morning clearing out wet fertilizer that had clogged the planter and—with it all—managed to finish planting this section exactly one month to the day ahead of last year's performance. Jack was touring Paris, France, at that very moment before traveling on to meet his hosts in Germany. Not bad for a Missouri farm boy!

# Chapter Fifteen

Late in June I rode along with Dennis in the combine to view one of the first fields being harvested. To my astonishment, I discovered that what looked from the road like a full rich field held virtually nothing. The excessive cold wet weather had taken its toll on this year's crop. Where the wheat was good it was exceptionally good but, unfortunately, these favored patches were few and far between. For the most part the wheat had frozen off and the ground was bare. In some places the heads looked fat and full but they contained no seed.

As a result of the year's bad wheat crop, dreadful rumors abounded. Unfortunately, most of them proved true. We heard of a man who drove a large farm truckload into town to sell. After the farmer was docked for moisture and weed content, the buyer said he would take the wheat if the farmer paid *him* three cents a bushel. As it wasn't even good enough to mix for feed, the farmer finally had to dump it in a ditch.

Most of the wheat in the fields around our home had been beaten flat by recent storms. In normal years Dennis can clear these fields in a few days. This year it took nearly two weeks, and still-twisted wheat stalks remained in the fields or clogged the thresher and had to be pulled out from beneath the spreader by hand.

And Dennis picked this moment to decide that he was not going to take no for an answer; I was finally going to learn to drive the combine.

I was again reluctant, but I hate to be a quitter, so I let him talk me around. This time he preset the reeler so that I would not have to keep raising and lowering it and could concentrate on the other aspects of the job. We adjusted the seat so I could drive and see. I made a practice run with Dennis sitting beside me, coaching. And then another, completed without mishap. Then he announced I was all set.

"I'm going into town with a load of wheat to see what I can sell it for," he said.

"No, you're not," I replied. "I can't drive this thing yet and you know how I hate to work in the field all alone."

"I won't even be gone an hour. Surely you can drive it for that long without screwing up."

That made me angry. I guess he knew it would—you can always get me to do something if you dare me.

So I headed back into the field and Dennis went off to town.

I managed fine for about twenty minutes. But my unpracticed eye failed to notice that the reeler was slipping inch by inch until it barely cleared the ground. The machine started making a horrible grinding noise and the header stopped turning. Nothing was being released from the back of the spreader. In less than an hour, I had screwed up. The threshing mechanism inside the combine had become clogged.

Forty-five minutes later Dennis informed me that because I had not made my turn into a row exactly square, the reeler had chopped the stalks in a sideways motion and that had caused the clogging. I was reprieved from trying to drive the combine, for a while.

At the end of the season, after all our work, we earned only a few thousand dollars from this wheat crop.

Jack returned from Germany, happy and more confident. After the necessary week at home to take care of his laundry, celebrate his birthday (and Kevin's real birthday), and Kevin's solitary First Communion, Jack was off to camp again.

Kevin learned to drive Gentle Ben, our 4040 John Deere,

which had been Jack's and my first tractor too. He was too short to reach the clutch, so he had to drive standing up, and for safety's sake Dennis only showed him how to engage first gear. But Kevin was so proud of his accomplishment. He also graduated to the duty of watering the smallest pigs and feeding sows in the farrowing room. He went around telling everyone he was learning how to take over Jack's jobs. And I guess maybe he is.

That summer Kevin wanted to go three places: the St. Louis Zoo, Six Flags Over Mid-America, and Wal-Mart. We were quite willing to take him up to the St. Louis Zoo in Forrest Park. He and Jack and I had been there once when Kevin was only four on a 4H Club trip, and we were surprised at all he remembered and was able to tell Dennis, who hadn't been there for many years. But Six Flags was a different story. It is a large amusement park with spectacular rides. Dennis had never been to it and had no desire to go. In fact, he accused me of making him go to punish him. The truth was that Kevin had been offered a free pass in school if he read for six hundred minutes in the course of six weeks and he had read so much that he had accumulated a much higher total. He would not sit still until we took him to Six Flags and I thought he'd earned it.

Right inside the gate we were confronted by Colossus, an eight-story-high Ferris wheel. The seats are really wire cages that tip over if they are unbalanced by the occupants. Dennis and Kevin enjoyed this ride immensely, particularly their view of me, eyes tightly shut, head bowed, clinging with both hands to the safety bar, trying to keep our cage upright.

Our next treat was Thunder River, where, after a half-hour wait in line, we got to take a four-and-a-half-minute ride from which we emerged absolutely drenched. I had to locate a restroom so I could remove my clothes and wring them out.

We went on to Buccaneer, a replica of a clipper ship that swings end to end, forty feet in either direction: Mid-America discovers seasickness. Feeling ill, we opted for the indoor, air-conditioned Chevy show and, while chewing Rolaids, viewed a racing film on a 3-D screen that had been taken by a camera attached to the first car. They had managed to convey all the delights of a ride on Ninja, which is the biggest roller coaster at Six Flags. It occurs to me that we have a funny definition of "amusement" in this country, but Kevin was happy.

Alas, Kevin did not get to Wal-Mart. But Dennis and I did, on our way home from *vacation*. We were determined to get away for a real vacation and, although it wasn't easy, we finally did it.

After I'd wrecked the Thunderbird, we decided to have it repaired and painted: I was not going on this vacation in a tractor or a pickup. But after our thirteen-year-old car went into the shop, they found a tiny pinhole rusted through on the right rear fender, and they wanted an additional $100 to fix it. Then it took more time than they had anticipated to grind the body smooth before the primer could be applied. A week after the car went into the shop we were told it would be ready in three days, maximum. Three days later, they informed us that the paint had lifted from one of the doors and it had to be completely redone. But they felt bad about all the delays, and promised the car for Friday. So I held my breath and went ahead making our reservations for a trip beginning on Saturday. And Friday evening, driven by Dennis accompanied by Kevin, the T-Bird made it home.

Our first trip in our shiny restored car was three miles down the road to drop Dennis at the 4455, which was in a field where he'd been discing weeds. We went so slowly over the county road to avoid having rocks thrown up, for fear of chipping the new paint, that, with the air conditioner blasting in the ninety-five-degree heat, the motor began to heat up. So we turned the air-conditioning off. Dennis transferred to the tractor, to bring it in, and I followed him back home in the car. But when I stopped to close a gate behind us, the car wouldn't start up again. Dennis, in the tractor, was oblivious to my situation and, never looking back, kept on driving.

After hopping about in the road for a while like an idiot, I walked down the road and found someone at the nearby winery who let me telephone. In a quarter of an hour Virgil arrived with jumper cables to rescue me.

When I got home, Kevin informed me that the mechanic at the shop had warned them that the battery was low.

Later that evening, we found that a front tire was flat: In the short time it had been back from the shop, the T-Bird had somehow picked up a screw. And the oil still had to be changed. And the engine had to be greased. Dennis wanted me to help out with this job but try as I might I couldn't seem to position the car over the ditch at the roadside well enough so that Dennis could slide beneath it to work.

The next morning we were wide awake by six-thirty. Dennis was still anxious to grease the car before we set off. But Kevin came back from feeding the sows to report that a sow with five three-week-old piglets had died during the night. So we had to delay everything else to pull the dead sow out of the building. Dennis has a grappling hook for the purpose, but this was a four-hundred-pound sow and it took the combined strength of Dennis, Kevin, and me to haul that sow out of the crate, down the aisle, and out the door. We buried her in our graveyard in the woods and Dennis placed the baby pigs with another sow. Then, after the routine feeding was finished, we had to move two sows into the farrowing building, one from the finishing building and one from the outdoor pens. To our surprise, the sow in the outdoor pen gave no trouble, but the other tore down every board, gate, and tin, protesting her move, and each of these items had to be repaired.

It was eleven-thirty before Dennis came in to shower and shave before driving to town to have the flat tire repaired. That gave me a chance to water my plants, indoor and out, and do the dishes, bathe, and finish the packing.

We set out at three in the afternoon for our four-hour journey south, so exhausted from the struggle to get away for five days of vacation that we really needed one.

Dennis and I are both big country-music fans so we had decided to head for Branson in the Missouri Ozarks, a town that is giving Nashville some competition as a country-music center.

We went to Silver Dollar City, saw performances by some great singers and musicians, and in four days encountered more people than we sometimes see in six months. Each night we returned to our motel, exhausted from being entertained, to attempt to sleep to the serenade of the clinking ice machine right outside our door.

The boys had no animals to display at the Gasconade County Fair so we went to the fairgrounds only as spectators just to see the antique tractor show and for the final night.

This would be Jack's senior year in high school. He was worried about his reception when he returned to school but there was a new principal and his attitude was entirely different. He even got a container of sugar to keep in his office, in case Jack needed it. Almost at once Jack had to start obtaining college admission information, registering for and taking entrance exams, and filling out scholarship applications. But above all else in importance was the passage of a major milestone—he got his own car.

And suddenly Kevin, who had grown six inches and put on thirty pounds in the last year, was a big boy. Thank God, Teenage Mutant Ninja Turtles were still his main interest, though I noted he spent a lot of time moussing his blond crewcut to make his hair stand on end. He told me that he was looking forward to school. He was bored at home, he said. All we did was work.

There was a milestone for me too. All my life I have been writing; for years this has been my ambition, always deferred to the prior claims of motherhood, domestic chores, and farm work. On August 31, 1991, I learned that an article I had written and sent to a national publication would be published.

They were going to pay me the grand sum of three dollars and sixty-three cents, but that was not the point. Dennis laughed, and Jack "thought I had lost it, totally," although Kevin hugged me, but for me, this was an achievement. Finally, after twenty years of working alone, I had crossed the line between being an aspiring but unpublished writer and seeing my thoughts and words in print, accorded value by someone else.

During the winter months I write at night after 9:00 P.M. Sometimes I work until two and three in the morning, glad of time without interruption.

Dennis insisted he could not sleep unless I was by his side, so we moved the desk with my typewriter and papers from the living room into the bedroom next to his side of the bed. Despite the hum of the typewriter, Dennis sleeps peacefully, knowing I am there. In planting season I carry pencils and paper into the fields, and I write in the few moments I can seize. I took my manuscript on vacation and wrote a few pages each night. Writing is as necessary to me as breathing. I am determined to keep going; I am determined to write; and I hope that someday I will be the author of a published book, but whether or not that happens, I will keep writing.

In spite of the results of my last attempt at driving the combine, Dennis was undaunted. He began to warn me that this was the week I was going to do it.

"Oh no, I ain't," was my reply, but he refused to listen.

Deep inside I relished the idea of finally conquering the monster machine. However, I was not going to admit this to Dennis.

The first combine I ever rode in was the old F2 Gleaner Dennis purchased right around the time of our marriage. It didn't have as many electronic controls and was somewhat smaller than our present F3 that I call Dino the Dinosaur but it still put me in awe of its sheer power and of the novel perspective on the world it grants you. The view from the cab of a combine is a grand one. You look way down over the tops of corn stalks and bean plants, watching as the reeler (or head) catches each stalk of corn, beans, or wheat and devours it, rolling it beneath the cab into the thresher. You get a sense of euphoria as the job is done, an enormous and gratifying sense of power under your control. Unfortunately, my latest misadventure in June did not give me much confidence in my ability to control this awesome beast.

"Shelling corn is much easier," Dennis assured me. "It's not nearly so particular."

I had a hard time believing this. There is nothing Dennis does about which he is not over-particular.

"Steering is simple," he went on.

"It's not steering between the rows that I'm worried about," I told him.

He could not imagine what my problem could be.

I mumbled my confession. "It's the auger."

He was completely puzzled.

"I'm not sure I can drive that monster right alongside the truck well enough so that I can get the auger in position over the truck bed. I'm afraid that I won't unload the corn into the center of the truck bed—what if it all ends up in a heap on the ground?"

"You don't have to position it perfectly over the center of the truck bed," Dennis growled. "Just get it in there."

"But what if I wreck it?"

"Even you couldn't do that," he snickered.

But I could envision myself running Dino through a fence or catching the corn-head points and wrapping them around a tree trunk and I was not reassured.

Nevertheless, Dennis was determined. He told Kevin he couldn't ride along in the cab of the combine because he didn't want me to be distracted, which offended Kevin but relieved me. Then he gave me a quick review of operating procedures and although it had been three months since my last ill-fated attempt, it all came back to me very quickly.

It was simpler to point the head through rows of corn than it had been through the wheat. I was even brave enough to try a round alone without Dennis beside me in the cab. And this time, unlike my attempt three years earlier, I not only stayed on course but didn't even tip the ground with the header.

Come Friday we were up early. I had to get the boys off to school and run some errands in town before I went to the field. Dennis and Virgil each took a truckload of corn from the previous day's harvest into the local elevator to sell. Then I met Dennis in the field for the big event—Mim masters Dino.

First, however, we found we had to replace a worn sprocket and chain; then we had to grease the header and steering mechanism and fuel up. Then, I ate a quick lunch while Dennis combined the first two rounds. I rode one round with him so he could show me the pattern of the rows to be combined. Right in the middle of this round, the rear elevator that brings the shelled kernels up into the bin became clogged. We emptied it out and Dennis readjusted it, assuring me that this was unlikely to happen again, even to me. Then Dennis took off to town to sell a truckload at the elevator and I was on my own.

I rode the rows without any problem. And to my own amazement, emptying the full bin into the trucks was easy. I used the tracks beside the trucks as parking guides and had no problems. I worked away, filling four truckloads, each containing three to three and one-half binsful.

I had dreaded going through one very wet sinkhole in the very middle of one field. When I could evade it no longer, I followed Dennis's instructions and just throttled down and steered straight through. The tail end of the combine fishtailed around a little bit, and fishtailing in a combine is an indescribable sensation. But it didn't get stuck. At last, I thought, I've done it. I've mastered the combine.

The next day, Saturday, back at the field, Dennis and Kevin made less than one full round when two of the main belts on the combine broke. In utter disgust at the waste of the day, we took off for the Deutze-Allis dealers in Morrison, which is about twenty miles away, to purchase new belts. Then we circled back through Hermann to buy some ready-made sandwiches and to tell the men waiting for Dennis at the elevator past their ordinary Saturday noonday closing that Dennis would not be bringing in any corn for sale that day after all so they needn't stay on for him.

It was four o'clock before the combine was ready. I climbed inside again, now looking forward to the chance to drive it.

As I filled a truck from the combine's bin, Dennis would drive it away to our private storage facility below our house. By eight o'clock I had filled up four truckloads.

It was getting dark. By this time the windows of the cab were dusty and dirty so they reflected light and movement rather than allowing me to see through them. Two huge eight-inch-in-diameter spotlights on the front of the cab illuminated the way before me, two lights shone down on the header—one on the platform outside the door, one on the bin—and two more lit up the rear of the machine. For about ten feet I was surrounded by light. All the rest was darkness. Driving a tractor after dark had given me an eerie feeling before, but that didn't compare with the strange sensation that engulfed me now as Dino slowly lumbered through the night, going up and down between the endless rows of corn.

In the darkness I had difficulty telling where to turn into the rows. And once I turned down a row of that long one-hundred-and-twenty-acre field, it was just Dino, me, and the corn—stalk after stalk after stalk. I couldn't see any stars, any shadows, any other movement of any kind now. The only sounds I could hear were the grinding of the motor and the crackling of the stalks as they were drawn between the points of the head and the ears were ripped off by the chains and sucked into the thresher. I felt as if when I reached the edge of that field of corn, the world might end too.

If the machine had stalled, I would have locked myself inside the cab and stayed there until noon the next day. Nothing would have induced me to climb out of the cab into that wilderness, that jungle of stalks of corn. But the combine didn't stall and I vanquished my personal monster that night. After that, driving Dino no longer held any terrors for me.

But there is one monster that every farmer fears that will not be banished: the threat of financial failure. When a farmer's business fails he often loses not only his ancestral home, but land that has been passed down from his father and grandfather, and even earlier generations.

My maternal grandparents raised a family of fifteen children on farms during the Great Depression of the 1930s. Both were in their eighties when they died. My aunts, uncles, and my mother have often spoken of the hardships they endured during those years. They had to move many times; often they lived on bacon-and-milk gravy (a mixture of grease and flour, with a bit of milk to give it a smooth white texture). The bacon was sugar-cured in the smokehouse out in back of the home and came from their own home-fattened and butchered hog. When they died, neither of my grandparents owned much more than the shirts on their backs.

Virgil had united the farms of his mother's parents and his father's parents, but he had grown up knowing the lean years too, battling to save the sparse profits of crop sales and taking jobs off the farm to make ends meet in order to save the land.

My stepfather's family had also farmed the same acreage generation after generation. When his father retired in the 1960s, Daddy began to farm his land but it was not enough. He had to run a construction business and farm on the side to earn enough to raise three children in some comfort. But he often spoke of his dream of retiring from construction to devote himself to the farm. The flood of 1986 destroyed his ancestral home. He sold the sand that buried his hay field to an asphalt-and-gravel company, and he can still farm forty acres or so, but the river took his retirement dream from him.

At Fricke Farms we have managed to survive the 1980s farm crisis by the skin of our teeth. We have been lucky: 1982, 1983, and 1985 were good years in which hog and crop prices yielded enough profit to enable us to stagger our debts and keep operating. But we lost almost all of our crops in 1986. Without crops for feed or cash with which to buy feed for the hogs, we were in deep trouble. We sold off hogs, and in short order our two-hundred-fifty-sow herd was reduced to eighty sows. (Of course, we still had a number of fattening hogs to keep us going). With the money from that sale we were able to make our interest payments and pay other outstanding debts. Government subsidies also helped us get through that devastating year. But it was painful to write checks to pay bills for seed, fertilizer, and crop sprays that were used to grow crops that we were never able to harvest.

Everything on a farm goes through time's cycle—lean and productive years follow one another. Despite the dry weather, 1987, 1988, and 1989 turned out to be exceptionally productive and even profitable. Then we were confronted by taxes.

This too seems to be a cycle, a vicious one. To avoid being wiped out by Uncle Sam, we have to convert income into return on capital. To make a capital investment, we need loans, and to get a loan you need collateral. Farmers ran into so much trouble in the 1980s because they counted on the prospect of bountiful crops to pay off the interest and principal of the loans they took out to buy tractors or put up new buildings, but rising interest rates swallowed up the crop profits before the principal could be paid off. Or, worse still, there were no profits. Some homes were mortgaged to meet operating costs, and then they were lost too.

Today it takes incredible managerial skill to stay on top of the market. American farm produce is sold worldwide, so our income is not necessarily determined by local weather conditions or the quality of the crops we raise or store, or by domestic demand. We are at the mercy of governmental policies, of whatever deals the government makes—or doesn't make—to sell to Russia, Japan, or South America. To survive, we have to know what the international picture looks like and where world trends are heading. And how many people do?

We've bought land and machinery since 1989. We manage

to pay our bills on schedule. Here on the farm we survive one season at a time.

Last year, 1992, was a very good one for us. So the disastrous flood of July 1993 seemed, somehow, even harder to take. I have lived within five miles of the Missouri River all my life. That it floods is a fact we live with—an event we expect to occur. But I have never before seen a flood in July. And this year's floods were the worst I've ever seen.

No one could remember a year in which there'd been this much rain. It began in November and by the time spring came the ground was already saturated. During the month of June we had thirteen inches of rain, about eight inches above normal, and in the first week of July another ten inches fell upon our farm.

We were still trying to plant soybeans on Monday, July fifth. The Gasconade River had risen but then fallen so we had decided to take our chances and we managed to plant several acres in the mud. Tuesday night the ceiling in the utility room gave around the chimney and rain leaked through and flooded our hall carpet. At the crack of dawn the next morning Dennis went out to inspect our other buildings and the hogs. He returned to report that the hogs and hog buildings were fine but Frene Creek had flooded, submerging his parents' driveway and washing away their foot bridge so they were stranded.

By 6 A.M. we were out only to find the county road underwater. Dennis went to get Goliath, our big tractor, to move it and the corn planter onto higher ground. The roads were impassable and he had to walk overland. The water kept rising. By the time Dennis returned to where I was waiting in the pickup the water had risen nearly eight inches. Dennis admitted to being a little nervous at that point; Goliath had begun to feel light enough to float as he drove it through four feet of water.

We had left our tractors, Ben and Godzuki, along with the new drill and disc, standing in a field on another section. We moved them, too, sinking in mud up to our calves crossing the high ground. We had managed to get about half of this three hundred acre field planted but corn that yesterday had been rich and green, standing nearly five feet high, was now invisible beneath the water.

Hermann was isolated by floodwater which spilled into the streets and cut off the highways. Jack, at my stepdad's in Treloar, called to say they expected to be flooded by nightfall. Rhineland was evacuated when the levee went.

We lost our cash crops and only a little corn survived. But our home, our equipment and the pigs survived. Things are going to be very tight for a while. We'll have to buy corn in order to feed the pigs or cut back our herd again. And we don't know how long it will take for the ground to dry out which will determine how much wheat we can sow in the fall.

But with a little help from God, we'll muddle through . . . perhaps better than we did before. Someday our home will be finished. Someday we will have paid for the land and machinery we have to buy to operate properly. Someday both Dennis and I will drive new vehicles. In the meantime, even though we can't promise Kevin a VCR, we will do our best to see that Jack gets his college education.

When Jack graduated from high school he was not only admitted to the college of his choice, but he also won several scholarships, the largest of which is the Missouri Bright Flight Scholarship, awarded to Missouri residents who achieve a score of thirty or higher on the ACT college entrance exam. It will provide Jack with two thousand dollars a year for five consecutive years, so long as he maintains at least a 3.0 grade average. Now he is enrolled at the University of Missouri. Jack lives with my parents much of the time so he can hold down a part-time job not far from where they live while driving eighty miles every week to the campus in Columbia, Missouri, to attend classes. And he made dean's list in his first semester.

It's Kevin whom we expect to take over the farm. He is the only one in the next generation to bear the Fricke name. It's a heavy burden to place upon a little boy's shoulders, and sometimes I have grave doubts. I urge Dennis not to push Kevin in this direction. Right now Dennis says if Kevin doesn't want to farm, he won't have to. I can't really picture him giving in so easily when the time comes. But maybe Kevin will want to take on the farm. Unlike Jack, he is a child who feels a strong need to conform and to please others.

But he appears to have a natural talent for art. He has a fabulous eye for detail. Unfortunately, however, he has inherited a degree of Dennis's color blindness. Kevin sees colors but has difficulty distinguishing between shades and hues. It's really too soon to tell what his decision will be.

Neither Dennis nor I attended college. Twenty years ago most of our high school classmates entered the work force right away. Only a few outstanding students, those who had well-to-do parents, or those who wanted to avoid the Vietnam War went to college. Dennis's best grades in high school were in agri-related subjects; he never had any interest in doing anything but operating the farm. I did well in English and history in high school but my mother's ambition for me was that I become a registered nurse. I was a poor science student and this was not a career that interested me. She insisted, but I refused. So after high school I drifted, doing factory work for a few years, until I began to teach in parochial school and found work that I loved.

I don't want to make the mistake my parents made by urging my children to go in a direction that appeals to me, not them. But I know very few people who are doing exactly the work they started doing after high school. Though farmers today need a lot of math and must study constantly and keep up with research, a college degree is not yet a prerequisite to the successful operation of a farm. But agri-business has expanded more rapidly and undergone more changes in the last forty years than in many centuries. It may be necessary for Kevin to have a college degree even to operate a farm business the size of ours, if that is his choice of occupation.

I don't want to see my sons struggle as we have had to, as our fathers and grandfathers did before us. But I am saddened by the disappearance of many of the small family farms that once existed in this region. I dread the thought of this happening to our farm.

I am not a politically active person. I believe in voting for the man rather than for the party. In local elections, I either know the people running or I know what others whom I respect think of them. When it comes to higher offices, it's a guessing game.

Few farmers thought that Ronald Reagan was a great president. President Bush seemed sincere, but he didn't seem able to get his plans acted on. Now as I write, we're in a recession. Probably the end of this decade will find us no better off financially than we were when we started out. I doubt that Clinton and Gore will help the farmer much. But personally, I don't give a damn how a politician spends his money or whom he sleeps with, so long as he does his job and keeps the promises that got him elected.

Farmers agree that it is time for a change, that the economy needs help. But no matter how bad things get, we remember that in the end there is always the land. Those who endure will be the ones who know the land and how to care for it. Not the bankers, the conglomerates, the corporations that lease their machinery by the day, or their animals by the head. The survivors will be the farmers who are not afraid to get on their hands and knees to feel the earth, who till the soil until it is fertile, who plant the seeds and nurture them, and then work and sweat to bring in the harvest.

We at Fricke Farms are determined to be amongst those who survive and thrive in the twenty-first century.